Springer Texts in Business and Economics

More information about this series at
http://www.springer.com/series/10099

Günter Knieps

Network Economics

Principles – Strategies – Competition Policy

 Springer

Günter Knieps
Institute for Transport Economics and
   Regional Policy
University of Freiburg
Freiburg
Germany

ISSN 2192-4333   ISSN 2192-4341 (electronic)
ISBN 978-3-319-11694-5   ISBN 978-3-319-11695-2 (eBook)
DOI 10.1007/978-3-319-11695-2
Springer Cham Heidelberg New York Dordrecht London

Library of Congress Control Number: 2014955668

© Springer International Publishing Switzerland 2015
This work is subject to copyright. All rights are reserved by the Publisher, whether the whole or part of the material is concerned, specifically the rights of translation, reprinting, reuse of illustrations, recitation, broadcasting, reproduction on microfilms or in any other physical way, and transmission or information storage and retrieval, electronic adaptation, computer software, or by similar or dissimilar methodology now known or hereafter developed. Exempted from this legal reservation are brief excerpts in connection with reviews or scholarly analysis or material supplied specifically for the purpose of being entered and executed on a computer system, for exclusive use by the purchaser of the work. Duplication of this publication or parts thereof is permitted only under the provisions of the Copyright Law of the Publisher's location, in its current version, and permission for use must always be obtained from Springer. Permissions for use may be obtained through RightsLink at the Copyright Clearance Center. Violations are liable to prosecution under the respective Copyright Law.
The use of general descriptive names, registered names, trademarks, service marks, etc. in this publication does not imply, even in the absence of a specific statement, that such names are exempt from the relevant protective laws and regulations and therefore free for general use.
While the advice and information in this book are believed to be true and accurate at the date of publication, neither the authors nor the editors nor the publisher can accept any legal responsibility for any errors or omissions that may be made. The publisher makes no warranty, express or implied, with respect to the material contained herein.

Printed on acid-free paper

Springer is part of Springer Science+Business Media (www.springer.com)

Preface

Networks are complex systems whose individual elements interact, and thus cannot be viewed in isolation. Both on the cost side and on the benefit side network-specific characteristics emerge. The building and operation of networks leads to economies of scale and scope, as can be seen in the energy, telecommunications and transport sectors. Externalities of network usage can be both positive (e.g. Internet traffic) and negative (e.g. traffic congestions). Furthermore, universal service objectives are traditionally very important, for instance, in the postal and telecommunications sectors.

Despite—or perhaps precisely because of—these network-specific characteristics, entrepreneurial competition strategies are of great importance in network sectors. In the liberalised network sectors, network services and infrastructure capacities constitute separate markets. Free market access for suppliers and network services is possible without owning a network infrastructure, but it requires access to the complementary network infrastructure capacities. In order to remain viable, network providers must develop competitive pricing and investment strategies. This is the only way to finance the necessary investments in networks, and it requires an entrepreneurial search for innovative pricing structures and an entrepreneurial determination of decision-relevant costs.

From a competition policy perspective there are a number of exciting questions: In which parts of networks is competition functioning? In which subparts of network industries is an abuse of market power to be expected? When and under which conditions must the owner of a network grant access to his facilities to other market players? What is the institutional division of labour between cartel authorities and regulatory agencies?

The aim of this book is to provide a systematic and comprehensive introduction to the field of network economics. The introductory chapter gives an overview of the central issues of network economics. Each of the following chapters explores one of these issues in depth by means of network economic analysis, illustrating theoretical results through practical case studies.

The book can either serve as a basis for an introduction to the field of network economics, or be used in lectures with a narrower thematic focus, for example, on transport, energy, telecommunications and the Internet. In addition it is targeted to all practitioners in network industries.

I would like to thank everybody who assisted and supported me in the writing of this book, both in its original German version of 2007 and in this revised and updated English version. For revision of individual chapters my thanks go to Martin Keller, Birgit Rosalowsky, Klaus Schallenberger, Volker Stocker, Marei Waidmann, Hans-Jörg Weiß and Patrique Wolfrum. In particular, I would like to thank Monika Steinert for copyediting and English language revision of the manuscript.

Freiburg   Günter Knieps
July 2014

List of Figures

Fig. 2.1	Deprival value	14
Fig. 2.2	Product-group specific common costs in the three-product case	26
Fig. 3.1	Socially optimal congestion fees	38
Fig. 3.2	Traffic density and traffic flow	39
Fig. 3.3	Congestion and hyper-congestion	39
Fig. 3.4	Capacity constraints	40
Fig. 3.5	Parallel paths	50
Fig. 3.6	Braess paradox	51
Fig. 3.7	Competition between parallel roads	53
Fig. 3.8	Uncongested travel on road A	54
Fig. 3.9	Congestion fees in a monopoly	56
Fig. 3.10	Loop flows (entry point 1)	62
Fig. 3.11	Loop flows (entry points 1 and 2)	62
Fig. 3.12	Reversal of merit order	66
Fig. 4.1	Peak load pricing under competition (firm peak case)	77
Fig. 4.2	Peak load pricing under competition (shifting peak case)	77
Fig. 4.3	Peak load pricing in a monopoly (firm peak case)	79
Fig. 4.4	Peak load pricing in a monopoly (shifting peak case)	79
Fig. 4.5	Optional two-part tariffs	80
Fig. 6.1	Utility effects of the transition to a new network	107
Fig. 7.1	Markets of decreasing density	125
Fig. 9.1	The regulatory triangle	159
Fig. 9.2	The Stigler/Peltzman model	163

List of Tables

Table 2.1	Economic depreciation	16
Table 2.2	Closed depreciation schedule	19
Table 2.3	Open depreciation schedule with one-time change of expectations	19
Table 2.4	Open depreciation schedule with two-time changes of expectations	20
Table 6.1	Standards as public goods, private goods and club goods	104
Table 8.1	Competition versus network-specific market power	137
Table 8.2	Monopolistic bottlenecks in selected network sectors	144

Contents

1 Introduction to Network Economics 1
  1.1 The System Character of Networks 1
  1.2 The Disaggregated Approach of Network Economics 2
      1.2.1 Network Levels 3
      1.2.2 Horizontal and Vertical Network Interconnection 3
  1.3 Economic Characteristics of Networks 4
      1.3.1 Network Externalities 4
      1.3.2 Economies of Scale and Economies of Scope 4
  1.4 Basic Questions of Network Economics 5
      1.4.1 The Role of Markets 5
      1.4.2 Decision-Relevant Costing 6
      1.4.3 Pricing Strategies in Networks 7
      1.4.4 Competition Policy and Market Power Regulation 7
      1.4.5 Universal Service Under Competition 8
      1.4.6 Compatibility Standards and Technical Regulatory
            Functions 9
  References 10

2 Decision-Relevant Costs 11
  2.1 Basic Principles of Determining Cost of Capital 11
      2.1.1 User Cost of Capital 11
      2.1.2 The Deprival Value Concept 13
      2.1.3 Implementation 17
  2.2 Decision-Relevant Cost Allocation 22
      2.2.1 Short-Run Versus Long-Run Marginal Costs 22
      2.2.2 Long-Run Incremental Costs 23
      2.2.3 Long-Run Incremental Costs Versus Long-Run Avoidable
            Costs 23
      2.2.4 The Traditional Concept of Overhead Costs 24
      2.2.5 Product-Group Specific Common Costs and Firm-Specific
            Common Costs 25
  2.3 Cost Strategies in Networks 27
      2.3.1 Network Evolution 27
      2.3.2 Strategies for Building a Network 27

	2.3.3	Decision-Relevant Costs of New Networks	29
	2.3.4	Long-Run Incremental Costs of Novel Network Services	30
2.4	Questions		31
References			31

3 Congestion Externalities ... 35
3.1 Local (Path-Based) Externalities ... 35
- 3.1.1 Congestion Externalities and Congestion Fees ... 35
- 3.1.2 Optimal Congestion Fees ... 36
- 3.1.3 Socially Optimal Congestion Fees and Investment Decisions ... 40
- 3.1.4 Efficient Congestion Fees and Financing Objectives ... 43
- 3.1.5 Congestion Externalities and Quality Differentiation in Infrastructure Networks ... 49
- 3.1.6 Congestion Fees in a Monopoly ... 54
- 3.1.7 Congestion Fees in Traffic Practice ... 57

3.2 System Network Externalities (in Electricity Transmission Networks) ... 61
- 3.2.1 Basic Characteristics of Electricity Transmission Networks ... 61
- 3.2.2 Wind Energy and Efficient Electricity Transmission Networks ... 68

3.3 Questions ... 70
References ... 70

4 Strategies for Price Differentiation ... 73
4.1 Basic Principles ... 73
- 4.1.1 Price Differentiation Through Peak Load Pricing ... 75
- 4.1.2 Price Differentiation Through Optional Two-Part Tariffs ... 79

4.2 Price Differentiation in Network Sectors ... 81
- 4.2.1 Price Differentiation for Network Services ... 81
- 4.2.2 Price Differentiation for Railway Tracks ... 82
- 4.2.3 Price Differentiation for Airport Slots ... 83

4.3 Questions ... 85
References ... 85

5 Auctions ... 87
5.1 Basic Principles ... 87
- 5.1.1 Elements of Auction Design ... 88
- 5.1.2 Fundamental Problems of Auction Theory ... 91

5.2 Auctions in Network Industries ... 93
- 5.2.1 Network-Specific Particularities ... 93

5.3 Disaggregated Invitations to Tender and Auctions in Network Sectors ... 96

		5.3.1 Invitations to Tender in Public Transport	96
		5.3.2 Auctions of Frequencies	97
	5.4	Questions	98
	References		99
6	**Compatibility Standards in Networks**		**101**
	6.1	Basic Elements	101
		6.1.1 Direct and Indirect Network Externalities	101
		6.1.2 Standards as Public Goods, Private Goods, and Club Goods	102
		6.1.3 Network Externalities Between Network Variety and the Search for New Technologies	104
		6.1.4 Standards for Goods, Complementary Components and Large Technical Systems	105
	6.2	The Coordination Problem	106
		6.2.1 Spontaneous Switching to a Superior Technology	106
		6.2.2 The Phenomenon of Critical Mass	108
		6.2.3 Path Dependency	109
	6.3	Conflicts of Interest	110
		6.3.1 Producers	110
		6.3.2 Consumers	111
	6.4	Standard-Setting Institutions	112
		6.4.1 Government Intervention	112
		6.4.2 Market Solutions	114
		6.4.3 Committee Solutions	115
	6.5	Standardisation of Technical Regulatory Functions	117
	6.6	Questions	118
	References		118
7	**Universal Service**		**121**
	7.1	Comprehensive Network Opening and Universal Service Objectives	121
		7.1.1 Services of General Economic Interest	122
		7.1.2 Defining the Scope of Non-profitable Universal Services	122
	7.2	The Instability of Internal Subsidisation Under Competition	123
	7.3	The Concept of the Universal Service Fund	125
	7.4	Universal Services in Telecommunications Markets	127
	7.5	Universal Services in Postal Markets	129
	7.6	Questions	130
	References		131
8	**Market Power Regulation**		**133**
	8.1	Localisation of Network-Specific Market Power	133
		8.1.1 Competition Versus Market Power	133
		8.1.2 Market Power Due to Economies of Scale?	134
		8.1.3 Network-Specific Market Power	135

	8.2	Disaggregated Identification of Competitive Potentials in Network Industries.	137
		8.2.1 Competition on the Network Service Level	137
		8.2.2 Competition on the Infrastructure Management Level	139
		8.2.3 Competition on the Network Infrastructure Level	141
		8.2.4 Monopolistic Bottlenecks on the Network Infrastructure Level	143
	8.3	Disaggregated Market Power Regulation	144
		8.3.1 Monopolistic Bottlenecks and the Concept of the Essential Facility	144
		8.3.2 Case Study: Newspaper Delivery Service	145
		8.3.3 Limiting Regulation to Monopolistic Bottlenecks	147
		8.3.4 Anticompetitive Price Structure Regulation	147
		8.3.5 Price Level Regulation of Access Tariffs	150
		8.3.6 Implementation of Price-Cap Regulation	153
	8.4	Questions	154
		References	154
9	**The Positive Theory of Regulation**		**157**
	9.1	Normative Versus Positive Theory of Regulation	157
	9.2	The Positive Theory of the Behaviour of Regulatory Agencies	158
		9.2.1 The Cornerstones of the Regulatory Process	158
		9.2.2 The Legal Framework of Regulation	158
		9.2.3 The Regulatory Agency's Discretionary Freedom of Action	160
		9.2.4 The Influence of Interest Groups	161
		9.2.5 The Disaggregated Regulatory Mandate	166
	9.3	Questions	167
		References	168
Sketch Solutions to the Questions			**171**
Index			**181**

Introduction to Network Economics

1.1 The System Character of Networks

The term network is used in quite diverse ways. The most general way of describing networks is by using the concepts and terminology of graph theory, a modern branch of mathematics. A graph consists of a set of nodes (vertices) which are connected to each other by links (edges) (cf. Diestel, 2005, pp. 1ff.). Graph theory provides an analytical framework for studying different network configurations. The terminology of graph theory has been adopted in various fields outside of mathematics, such as sociology, engineering, urban economics and regional planning. Thus sociologists characterise the intricate immaterial systems of relationships between human beings as social networks, defining single individuals (or organisations, respectively) as nodes and interrelationships as edges. From this perspective, economic networks are regarded as special cases in order to better understand the influence of social relations on market transactions (e.g. Zuckerman, 2003).

Engineers define networks as connections between different points, where the connections function as paths of transmission. Consequently, they differentiate between distribution networks and interactive networks. The function of the former is to provide consumers with water, gas, electricity, cable TV etc., and they are frequently structured like a tree. The purpose of the latter is the exchange of goods, information etc., e.g. road and railway networks, as well as telecommunications networks; they are star-shaped or circular or frequently intermeshed.

In order to understand the connective element in these types of networks, graph theory is again a suitable starting point. In traditional telecommunications networks, location-based end customer and switching facilities constitute the nodes, while all types of transmission facilities can be viewed as edges. In traffic networks, roads or railway tracks constitute the edges, while road crossings or railway stations are the nodes. Concerning the Internet, servers are regarded as nodes and the physical infrastructure between these servers (e.g. fibre optic cable) as the connection (edge). In this sense, transportation networks can be represented with the help of graph theory (e.g. Rodrigue, Comtois, & Slack, 2006).

One implication of the graph theory perspective is that networks have to be viewed as complex systems. The relevant elements (nodes, edges) cannot be examined in isolation, but must be studied in relation to the other network elements. Thus, for example, all edges (e.g. roads) leading to one node must be included in the analysis. The choice of connections is not an isolated but an interdependent process. The interaction of elements within a complex system simultaneously reflects the organisational principles on which the network is based and which enable it to develop and adapt to changes in the environment (cf. Spulber & Yoo, 2005, p. 1694).

The system character of networks implies that the individual elements interact and thus cannot be viewed in isolation. This system character is also the starting point of an interdisciplinary research approach which has become known as the Large Technical Systems approach (e.g. Hughes, 1983). The focus of this approach is on network industries such as electricity, telecommunications, transport, gas, water supply and sewage. Objects of study are the decisions made by firms, the effects of governmental regulatory measures on network sectors, and the extensive interaction of these systems with society as a whole. Particular emphasis is put on the dominance of hierarchical organisations and the strong influence of the state. Networks are constructs built by individual agents or by groups of agents for specific objectives. Thus, special importance is placed on the builders of networks (engineers, managers, financiers etc.) who develop, construct, and maintain technical systems (e.g. Mayntz & Hughes, 1988).

1.2 The Disaggregated Approach of Network Economics

From both graph theory and the Large Technical Systems approach it follows that a fragmentation of networks into individual components with different decision makers does not sufficiently take into account the character of networks, because essential system interdependencies would be neglected. However, this does not mean that all decisions within one network have to be made by one person or one group. The closed network where all nodes and connections are under the decision authority of a single network provider is replaced by the concept of open network access, where free market entry of network providers is the competition economic point of reference. Constructive planning processes are to be differentiated from the development of networks over time governed by path-dependency as well as evolutionary search processes.[1]

---

[1] For competition as a search procedure see von Hayek, 1968/2002.

1.2.1 Network Levels

In the case of physical networks, a broad distinction can be drawn between network infrastructures (tracks, airports etc.) and network services (railway transport, air traffic, etc.). Although network services and network infrastructures are complementary to each other, they constitute different network levels which can (with the exception of the necessary compatibility and security standards) be built and operated independently of each other. Therefore in the terminology of graph theory, vertices and edges belonging to the infrastructure level should not be mixed with those of the service level. From the perspective of the Large Technical Systems approach, this constitutes vertical disintegration of large technical systems.

Based on a disaggregated examination of the value chains in network sectors, end customer markets for network services can be differentiated from upstream markets for infrastructure capacities. Free market entry for service providers who do not own network infrastructure is possible; this does, however, require non-discriminatory access to the complementary network infrastructure capacities.

For the analysis of network economic problems the following classification of complementary network levels is useful:

Level 1: network services (e.g. air transport, telecommunications services, generation and retail of electricity)
Level 2: infrastructure management (e.g. air traffic control, railway traffic control)
Level 3: network infrastructures (e.g. railway tracks, airports, telecommunications networks)
Level 4: public resources, on the basis of which network infrastructures and infrastructure management can be built (e.g. land, air, space, water).

This classification system differentiates between the markets for network services (Level 1), the markets for infrastructure management and network infrastructure capacities (Levels 2 and 3), and the upstream markets for public resources (Level 4). The significance of the economic characteristics of networks can vary substantially between network levels.

1.2.2 Horizontal and Vertical Network Interconnection

Due to the market opening of networks, issues of organisation and network interconnection become increasingly more important. On principle, networks can be subdivided into different, complementary levels, the separation of which is technically and organisationally feasible. In the process, horizontal as well as vertical interconnection problems occur.

Horizontal interconnection problems can occur on the level of network services as well as the level of network infrastructure. For instance, the coordination between different airlines makes it possible to exhaust economies of scope in larger, interconnected air traffic networks by a joint aircraft deployment policy etc. There is also a considerable potential for coordination and cooperation in the realm of network infrastructures.

The provision of network services is only possible when access to the complementary network infrastructures is guaranteed at the same time. Symmetric access to these facilities is required for all active and potential suppliers of network services.

1.3 Economic Characteristics of Networks

1.3.1 Network Externalities

One must differentiate between positive and negative network externalities. Positive network externalities can be illustrated using the example of a telecommunications network. The larger the number of customers connected to a specific telecommunications network, the larger the benefit of network connection, because with an increasing number of users the possibilities of reaching and being reached by other customers also increases. Negative network externalities take the form of congestion costs resulting from the utilisation of network infrastructures. An example for this is traffic congestion on the roads. Motorists typically neglect the negative effects (e.g. delays, and thus longer travel times) that an additional trip at a certain point in time can have on the other road users.

The third relevant category of network externalities is system network externalities. System network externalities occur due to physical-technological characteristics of networks. They are of central importance in electricity transmission and distribution networks. In an electricity network it is not possible to transmit electricity between an entry and an exit point without at the same time affecting the opportunity costs of network utilisation in all nodes of the other parts of the network. System network externalities can be either positive or negative.

1.3.2 Economies of Scale and Economies of Scope

When examining the cost side of networks, economies of scale and economies of scope are of central importance. Economies of scale (increasing returns to scale) exist, if a proportional increase of all input factors causes a disproportionally high increase of all output components. If there are economies of scope, it is less costly for a single firm to produce all product lines together than it would be for different firms to specialise in producing separate product lines. These cost advantages may be based on joint production (e.g. the joint utilisation of inputs) and/or on advantages in the distribution of products. Economies of scale and economies of scope may be relevant both for the building of network infrastructures and for the provision of network services. Economies of scale and economies of scope can lead to a natural monopoly, where a single provider can supply the relevant market at lower cost than several providers. From the perspective of the disaggregated approach of network economics it is important to differentiate between natural monopolies on the infrastructure level and natural monopolies on the network services level.

Let us consider a scenario where a water network is being planned. First of all, there are economies of scale with regard to the circumference of the water pipe. The larger the number of houses to be served along a street, the larger the diameter of the water pipe and thus the lower the cost per house connected to the pipe, because when the diameter of the pipe is increased, the volume of water grows faster than the circumference of the pipe, which ultimately determines the cost. While economies of scale are only relevant for individual given pipes between two nodes (edges), economies of scope occur when different pipes are connected to each other over space. This leads to the problem of weighing the advantages of exhausting the economies of scale of the pipes against the additional pipeline length required for doing so. For instance, it might be less costly to first build a large service pipe to a joint distribution centre and serve individual pipes from there. Finally, given (partly) stochastic demand over time, certain smoothing effects occur, which also reduce the average network capacity required when the number of users increases.

Economies of scale and economies of scope can also be found in other network infrastructures, e.g. energy networks, telecommunications networks, road and railway track networks. Obviously, these advantages occur with tree networks, ring networks, interactive star networks (such as in telecommunications) as well as meshed networks. However, economies of scale and scope can also be relevant on the level of network services. One example for this is the formation of hubs in air traffic. Further examples are postal delivery services or the provision of public transport services.

1.4 Basic Questions of Network Economics

1.4.1 The Role of Markets

1.4.1.1 Markets for Network Services

At first sight the transport of people and goods on roads, railway tracks, waterways, as well as in air corridors, and the provision of electricity, telecommunications and Internet services constitute very different markets. What they have in common, however, is that they are all network services, which require the use of network infrastructures for their provision.

The provision of network services is a private good, characterised by the possibility of exclusion as well as rivalry with regard to consumption. Due to the fact that services cannot be stored, shifts of demand over time can lead to capacities—e.g. of airplanes or trains—not always being used to full capacity. In order to counter this problem, on the one hand there is the possibility of adjusting demand to the capacity made available by choosing suitable dimensions for transport vehicles, power plants etc., and by optimising their use; on the other hand, the utilisation of suitable pricing instruments (in particular peak load pricing) enables the smoothing of fluctuations in utilisation over time.

1.4.1.2 Markets for Network Infrastructure Capacities

In the past the building and operating of network infrastructures, in particular in the transport sector, has been regarded as a government responsibility, because it concerns public goods that ought to be financed by public expenditure. Although Smith (1776), the pioneer of classic liberalism, viewed competition between the individuals active in the market as the crucial control mechanism, he regarded the provision of infrastructure (roads, bridges, canals, ports, water supply) as a public responsibility.

If indivisibility in the building of network infrastructures leads to a complete absence of rivalry in their use, charging user fees in order to allocate capacities is not expedient. Thus the responsibility for setting the socially desirable investment level, as well as guaranteeing its financing lies with the state. In 1919 the Swedish economist Erik Lindahl (1919, pp. 85–98) already suggested the following solution: Every citizen should contribute an amount corresponding to his or her marginal utility from the public investment; its use, on the other hand, should be free of charge. Even from this argument, however, it does not follow that the subsidisation of infrastructures is justified in all circumstances. If social welfare (the sum of consumer and producer surplus) generated by an infrastructure is lower than the decision-relevant cost, then the question arises as to the economic justification of a further upgrading or a downgrading of network infrastructures.

When demand for network services increases, the (derived) demand for network infrastructure capacity increases simultaneously. Thus the transport volume, for example, has grown substantially over the last decades, which has led to significant scarcity problems on many airports and motorways. But at peak periods rail track networks, electricity networks and telecommunications networks are similarly working at their fullest capacity. Market-compatible access charges must guarantee a non-discriminatory access to network infrastructures; in addition they serve the function of an optimal allocation of existing network capacities and, if possible, they should cover the cost of the network infrastructure.

1.4.2 Decision-Relevant Costing

In network sectors that provide water, electricity, telecommunications services, etc. the conceptual evaluation of decision-relevant costs is of particular importance (cf. Knieps, 2001, pp. 288ff.). In comparison to other industries, the building of network infrastructures requires a particularly large capital expenditure. An economically well-founded calculation of the cost of capital for each period is therefore of special relevance for the evaluation of decision-relevant costs. Network operators typically are multi-product firms. Simultaneously serving adjoining houses or streets generates economies of scope. When working with concepts of decision-relevant costs, a differentiation must be made between

- long-run incremental costs (in order to answer the question whether the provision of certain network infrastructure capacities or network services should be discontinued or, conversely, its extent increased)
- stand-alone costs of products or product groups, respectively (in order to answer the question whether the building of alternative partial networks is cost-covering)
- total cost (in order to determine the viability of the active network operator).

Concepts of decision-relevant costs are relevant for the level of network services, as well as the level of network infrastructure (cf. Chap. 2).

In order to ensure the viability of network providers, access tariffs have to be designed in such a way that the necessary investments in the networks infrastructure can be made. Optimal network access charges equal to the opportunity costs of network utilisation at a given level of infrastructure quality serve the function of allocating existing capacity. As optimal network access charges contribute to the financing of infrastructure, opportunity costs constitute a suitable link between the pricing of network utilisation and the recovery of the total cost of the network. Thus network access charges that take scarcity into account help achieve the financing objective.

1.4.3 Pricing Strategies in Networks

Entrepreneurial pricing strategies for network access thus have to aim at taking the opportunity costs of network utilisation into account as fully as possible. The value of the best possible alternative use of a network capacity determines the level of these opportunity costs. Optimal network access charges equal to the opportunity costs of network utilisation at a given level of infrastructure quality lead to optimal allocation of existing capacity. A multitude of allocation mechanisms can be used to solve problems of scarcity in network infrastructures, including congestion externalities. The intramodal analysis of congestion fees and related investment guidelines have a long tradition in transportation economics. Mohring and Harwitz (1962) already provided theoretical insights into the relation between socially optimal highway tolls (based on congestion externalities) and socially optimal investment levels. Relevant entrepreneurial strategies for congestion charges and price differentiation as well as auctions in network industries are provided in Chaps. 3–5.

1.4.4 Competition Policy and Market Power Regulation

The market opening of network sectors has expanded competition policy's sphere of competency substantially. Deregulation means that sector-specific competition policy exemptions are abolished, while general competition law becomes the basic point of reference. Only if the instruments of general competition policy are not sufficient should sector-specific market power regulation be introduced.

The competition laws in different countries not only focus on a per se prohibition of cartel agreements but also enable a control over possible abuses of dominant positions as well as merger control. In contrast to government regulations for the ex ante disciplining of market power which are applied generally for a given individual network sector, the rules of cartel laws regarding abuse of market power are predominantly applied on a case-by-case ex post basis.

On deregulated markets competitive strategic behaviour (e.g. mergers, alliances, price differentiation etc.) plays an important role. This is a natural consequence of the freedom created by deregulation and an expression of competitive potentials. The transformation of legal monopolies into competitively organised markets made it necessary to consider the application of competition policy measures, e.g. for combating anticompetitive collusion between competitors (which restricts competition) or exclusion of competitors (which prevents competition). As on all markets, it is the competition agency's responsibility to detect actual distortions of competition and the accompanying market power. In doing so, they face the difficult task of carefully weighing the economic damage resulting from an unjustified intervention against the economic damage resulting from an unjustified non-intervention.

The transformation of traditional legal monopolies into a competitive economy confronts competition policy with completely new questions: For instance, should an airline be forced to make its computer reservation system accessible to its competitors? How should frequent flyer programmes be assessed from a competition policy perspective? How can predatory pricing by infrastructure providers be prevented, if they are simultaneously active as service providers? These questions illustrate the broad spectrum of the manifold challenges faced by competition policy in the wake of market opening.

A suitable normative economic reference model for defining the sector-specific ex ante measures required for disciplining market power in network industries must be capable of examining the basic characteristics of networks (economies of scale and scope, network externalities), without automatically equating them with market power. What is necessary, then, is the localisation of network-specific market power and its disciplining with the help of suitable regulatory instruments (cf. Chap. 8). In the context of the positive theory of regulation the emergence, the transformation and abolition, as well as the institutional implementation of sector-specific regulation are being analysed (cf. Chap. 9).

1.4.5 Universal Service Under Competition

Universal service includes the provision of certain services at a politically desired tariff either on the infrastructure level or on the network services level. In the days of legal barriers to entry, it was the responsibility of the legally protected network provider to supply non-profitable universal services. Thus Deutsche Post, the former German postal services monopolist, was obligated to deliver letters in every region of Germany (both city and rural areas) at a socially desired uniform tariff. In opened network sectors the questions of when, where, to what extent, and at what price universal services are to be provided have to be answered explicitly in the context of a political decision-making process. In the financing of universal services as well as their subscription, individual businesses may neither be favoured nor discriminated against. The possibilities of universal services under competition will be analysed and illustrated in particular on the basis of the example of telecommunications and the markets for postal services (cf. Chap. 7). In particular, the liberalisation of network industries should not be hampered by the implementation of universal service goals.

1.4.6 Compatibility Standards and Technical Regulatory Functions

From positive network externalities it follows that the individual benefit from joining a network may possibly be smaller than the global benefit for all users. The crucial question is whether enough users will join to guarantee that operating the network will be cost-covering. For any given network, this is the problem of critical mass of network users. It is relevant for building a network as well as for switching from one network technology to another. In addition, there is the problem of network fragmentation, if several similar networks reach critical mass, but, due to lack of compatibility, the advantages of network externalities cannot be fully exhausted. The larger the number of individuals using the same network the larger the benefit. But this principle is limited by the heterogeneity of individual preferences for different technologies (network variety). Under certain circumstances it may be more beneficial for the group of all economic entities as a whole to split into several network islands than to build one large unified network. If compatibility cannot be achieved at all, or only at prohibitively high cost, the problem of the conflict between network externalities and network variety arises (cf. Farrell & Saloner, 1986). Network externalities make it expedient to choose network technologies that are as compatible as possible.

In network industries it is necessary to address the standardisation problem in a disaggregated way (cf. Blankart & Knieps, 1993). The starting point of technical regulatory functions (e.g. postal code systems, telephone number administration, land registers) are problems of coordination and allocation that precede the provision of network services and the building of network infrastructures. Technical regulatory functions can be relevant at every network level. They are particularly relevant for infrastructure management. A clear-cut geographic delimiting of the monitoring authority for railway traffic control or air traffic control constitutes a technical regulatory function. In contrast, the actual control functions in the context of capacity and safety management should be allocated to the level of infrastructure management and can be periodically contracted out (cf. Knieps, 2013).

Therefore the setting of standards for ensuring the compatibility of different technologies and enabling the interoperability between different parts of networks is very important in network industries (cf. Chap. 6).

References

Blankart, C. B., & Knieps, G. (1993). State and standards. *Public Choice, 77*, 39–52.
Diestel, R. (2005). *Graph theory* (3rd ed.). Heidelberg: Springer.
Farrell, J., & Saloner, G. (1986). Standardization and variety. *Economic Letters, 20*, 71–74.
Hayek, F. A. von (1968/2002). Competition as a discovery procedure. *The Quarterly Journal of Austrian Economics, 5(3)*, 9–23.
Hughes, T. (1983). *Networks of power: Electrification in Western society 1880–1930*. Baltimore: Johns Hopkins University Press.
Knieps, G. (2001). The economics of network industries. In G. Debreu, W. Neuefeind, & W. Trockel (Eds.), *Economic essays – A Festschrift for Werner Hildenbrand* (pp. 275–293). Berlin: Springer.
Knieps, G. (2013). Competition and the railroads: A European perspective. *Journal of Competition Law & Economics, 9(1)*, 153–169 (first published online February 6, 2013).
Lindahl, E. (1919). *Die Gerechtigkeit der Besteuerung*. Lund: Gleerupska Universitets-Bokhandeln.
Mayntz, R., & Hughes, T. P. (Eds.) (1988). *The development of large technical systems*. Frankfurt: Campus Verlag.
Mohring, H., & Harwitz, M. (1962). *Highway benefits: An analytical framework*. Evanston, IL: Northwestern University Press.
Rodrigue, J.-P., Comtois, C., & Slack, B. (2006). *The geography of transport systems*. London: Routledge.
Smith, A. (1776). An inquiry into the nature and causes of the wealth of nations. London: Methuen
Spulber, D., & Yoo, C. (2005). On the regulation of networks as complex systems: A graph theory approach. *Northwestern University Law Review, 99*, 1687–1722.
Zuckerman, E. (2003). On networks and markets by Rauch & Casella (Eds.). *Journal of Economic Literature, XLI*, 545–565.

Decision-Relevant Costs 2

2.1 Basic Principles of Determining Cost of Capital

Compared to other industries, the building of network infrastructure requires a particularly large capital expenditure. An economically well-founded periodisation of user cost of capital is therefore of special relevance for the calculation of decision-relevant costs. Concepts of decision-relevant costs are relevant for the level of network services, as well as the level of network infrastructure.

2.1.1 User Cost of Capital

Traditional production and cost theory analyses the supply side of markets, based on the technological conditions of production and the conditions of supply for production factors. If production function and factor supply are given, minimal cost combinations for each quantity of alternative output levels can be derived. Besides short-lived production factors acquired on the input markets and used in the production process over one period (e.g. energy, material, labour), more long-lived capital assets (e.g. machines) are also utilised in the production process. The derivation of the output isoquant for determining the optimal combination of capital and labour therefore presupposes a periodisation of user cost of capital. User cost of capital includes the decline in value (economic depreciation) and the cost of capital of investment (cf. e.g. Enke, 1962; Turvey, 1969, p. 286). The period-based evaluation of user cost of capital (economic depreciation plus interest) built an important branch of cost theory, which is indispensable for the understanding of decision-relevant costs.

Turvey (1969, pp. 288ff.) characterises long-run marginal costs as the change of cost caused by a permanent increase in output and the accompanying capacity expansion. The long-run marginal costs in one period consist of the user cost of capital and the marginal costs of production. Economic depreciation has the objective of measuring the loss of value of a facility (wear and tear, risk of technical obsolescence, price and/or value shifts of the facility) during a given period of time.

The basic principle of economic depreciation dates back to Hotelling (1925). Machines constitute relatively long-lived capital goods the value of which is determined by the units of output they produce. The basic principle of economic depreciation is to measure the decline in value a facility experiences during a given period of time. The fundamental principle here is measuring the value of a facility at different points in time during its entire economic life. It is assumed that the value of a facility at a given point in time consist of the balance of the utilisation potential of its remaining lifespan. Thus economic depreciation is the difference between the replacement cost at the beginning and at the end of the period. Based on the assumption that the facility is working at full capacity at all times, the average decline in value per output produced can be derived. The hypothetical annual rental value of a machine is equal to the sum of economic depreciation plus interest.

When determining decision-relevant depreciation it is important to observe the following principles (cf. Knieps, Küpper, & Langen, 2001, pp. 760ff.):

- Capital Theoretical Profit Neutrality
  Depreciation should not include profit-based components and be profit neutral. Based on the superordinate long-time profit objective, this requirement can be characterised as the principle of capital theoretical profit neutrality. It conveys that depreciation is a cost issue.
- Market Reference
  Neglecting to take market developments into account when planning decision-relevant costs can lead to wrong decisions and management errors. This has given rise to the principle of market reference which demands, above all, that the development of prices and technology be closely observed. As regards the prices of production inputs, the supply market is of primary importance, but the prices on the sales market are also relevant.

  Thus it seems practical for a firm to distribute the depreciation of a facility during its economic life over different periods in such a way that even in case of an expected price decrease no scheduled losses occur. For economic reasons, a firm will not make an investment, if, given expected changes in (supply and sales) prices, the investment will incur a loss.
- Forward-Looking Determination of User Cost of Capital
  As decisions are always targeted for the future, user cost of capital always constitutes a forward-looking concept. Thus the determination of economic depreciation schedules cannot function without assumptions concerning relevant events in future time periods. First the question needs to be examined whether all relevant decision variables and constraints can be expected to be known at the beginning of the planning period (closed event space), and thus no unforeseen events will occur. In particular, it is assumed that no unforeseen changes in prices or technology will occur. In such a stationary world, it is possible to derive "closed" depreciation schedules only once at the beginning of the planning

- Costing versus Pricing

  Consider a long lived capital good with an economic life of several periods. If capacity is exhausted in each period, it is possible to distribute capacity costs over different periods, taking into account the demand curve (cf. Hotelling, 1925). However, the problem of period-based determination of depreciation can also be solved for a shifting degree of capacity utilisation over time, using the analytical instruments of peak load pricing (cf. Littlechild, 1970; Baumol, 1971). The higher the aggregated willingness to pay in a given time period, the higher the share this period contributes to the covering of total capacity costs.

horizon, because at this point in time reliable (rational) expectations with regard to future events exist.

The principle of determining depreciation on the basis of capacity utilisation should not lead to the fallacy of blurring the border between the separate areas of costing and pricing. Decision-relevant cost concepts are based on the differences in capacity utilisation at different periods that can be predicted at the time the investment decision is made, because it is only these differences that can influence the dimensioning of the facility. Rivalry in usage and assumptions regarding scarcity can therefore only refer to average capacity utilisation during the period in question. Pricing, on the other hand, is short-term, so that it can react in a sophisticated manner to short-term changes in demand, even when investment decisions cannot be altered (cf. Vickrey, 1985, p. 1333). Peak load tariffs can vary considerably, even over the course of a single day, depending on fluctuations of demand at different times of day. This enables a fine-tuning of fluctuations in capacity utilisation over the day resulting in a more even utilisation of the capacity (cf. Sect. 4.1.1). In contrast, adapting capacity to changing demand conditions constitutes a more long-term problem. Thus, in contrast to the assumption of basic microeconomic models that the long-run cost function will adapt to capacity requirements at infinite speed, the upgrading of facilities requires a non-negligible amount of time.

2.1.2 The Deprival Value Concept

The deprival value concept defines the value of a facility by means of the opportunity costs arising if this facility would no longer be available within an enterprise.[1] This also takes into account the possibility that it can be worthwhile to sell a facility during its economic life or not to replace it.[2]

---

[1] For an overview, cf. Bell and Peasnell (1997, pp. 122f.), Solomons (1966, p. 125), Atkinson and Scott (1982, pp. 20ff.).

[2] Depending on the context, a facility may consist of a network component or of a bundle of network components.

Fig. 2.1 Deprival value

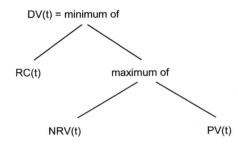

At a given point in time (the beginning of period t) the deprival value $DV(t)$ is determined as follows:

$$DV(t) = \min[RC(t), \max(NRV(t), PV(t))] \qquad (2.1)$$

whereby
$RC(t)$ denotes the replacement cost at the beginning of period t
$NRV(t)$ denotes the net realisable value at the beginning of period t and
$PV(t)$ denotes the net present value at the beginning of period t

The deprival value is illustrated by Fig. 2.1 (cf. Bell & Peasnell, 1997, p. 126). Three different cases have to be differentiated.

Case 1
Deprival Value = replacement cost

$$DV(t) = RC(t) \quad t = 1, \ldots, T \qquad (2.2)$$

If $PV(t) > RC(t) > NRV(t)$, $t = 1, \ldots, T$, it is profitable to buy a machine and to replace it if it fails. This also holds, if $PV(t) > NRV(t) \geq RC(t)$. A usage value of the machine equal to its net present value of future cash flows is the crucial factor for the machine being used within an enterprise during its entire economic life, because the usage value is higher than the net realisable value and not lower than the replacement cost.

Case 2
Deprival Value = net realisable value

$$DV(n) = NRV(n) \quad n \geq t \qquad (2.3)$$

if $RC(n) > NRV(n) > PV(n)$, $n \geq t$, the net realisable value represents the machine's opportunity costs. It is profitable to sell the machine at time t, because the net realisable value is higher than the net present value. Replacing the machine at replacement cost is not profitable, because the net present value is lower than the replacement cost.

Case 3
Deprival Value = net present value

$$DV(n) = PV(n) \quad n \geq t \qquad (2.4)$$

If $RC(n) > PV(n) > NRV(n)$, $n \geq t$ the net present value represents the opportunity costs caused by the elimination of the machine. Replacing or selling the machine is not profitable. Once purchased, the machine will be used until the end of its economic life. The question arises, whether the conditions of capital theoretical profit neutrality are fulfilled, even though deprival value is equal to net realisable value (case 1) or net present value (case 2). Under stationary conditions this is ensured, because the machine would not have been purchased otherwise. This will be shown in the following for case 3 (disregarding the option of selling the machine).[3]

By means of the following example, based on Wright (1968, pp. 225f.), it can be shown that an expected fall in demand for the output produced by the machine will, from a specific point in time onwards, result in a net present value which is lower than the replacement cost. If this is permanently the case, it is, from a specific point in time onwards, no longer profitable to purchase a new machine. In order to simplify the scenario, interest will be assumed to be 0 (cf. Table 2.1).

We will examine the example of a machine which produces 10 units of a non-storable product in each time period and is replaced after 2 years. At the beginning of each year, a new variant of the machine, functionally equivalent to the old machine, comes on the market. In year 1, there is one machine which is new. In year 2, one additional machine is purchased; the machine from year 1 is now old, but functionally as good as the new machine. Economic depreciation for each unit produced in period 2 is therefore valued equally, no matter if the unit is produced by the old or the new machine. The value of the utilisation potential of the more expensive old machine in period 2 is therefore equal to the value of the utilisation potential of the cheaper new machine in the same period. The replacement cost of the old machine must therefore be equal to the depreciation of the new machine (as long as the net present value is higher than the replacement cost). Consequently, the depreciation for the new and for the old machine is identical. The same argument holds for all following years, during which a new machine is purchased (as well as for the last period where only an old machine is used).

In the following example, illustrated in Table 2.1, it is shown that in periods 1–4 a new machine is purchased. In periods 5–7 no machines are purchased as they would lead to deficits. Hence, in period 5 only the old machine is used. In periods of active production, economic depreciation according to the replacement cost holds for the new and for the old machines in equal measure.

Declining product prices cause net present values in period 6 and 7 to fall below replacement cost. For the hypothetical situation in which a machine is purchased at

---

[3] An analogous argument holds for Case 2.

Table 2.1 Economic depreciation

Year	1	2	3	4	5	6	7
Price of the new machine $PP(t)$	100	70	50	40	40	40	40
Replacement cost $RC(t)$		40	30	20	20	20	20
net present value $PV(t)$		40	30	20	20	10	0
economic depreciation $d(t)$	60	40	30	20	20(30)	0(10)	0
labour costs	1	1	1	1	1	1	1
product price p_t	7	5	4	3	3	2	1

the beginning of period 5, the corresponding economic depreciation for periods 5 and 6 is given in brackets.

Disregarding the option of selling the machine, the deprival value is equal to the minimum of the difference between replacement cost and net present value.

$$DV(t) = \min\left[RC(t), PV(t)\right] \qquad (2.5)$$

The economic depreciation of a facility with an economic life of 2 periods is thus:

$$d(t) = PP(t) - DV(t+1), \qquad (2.6)$$

where $PP(t)$ denotes the price of the new machine (purchase price) at the beginning of period t.

In determining the replacement cost, the question whether the costs thus incurred can be covered by corresponding net present values of product prices is not examined. As the replacement cost in period 4 is 20, economic depreciation of the machine in period 3 is 30, because the original price was 50. Along these lines, economic depreciation in period 2 is 40 and in period 1 it is 60.

Product prices in periods 6 and 7 are assumed exogenously, because of intense competition on the output market $p_6 = 2$ and $p_7 = 1$. Therefore, in periods 6 and 7 the net present value is lower than the replacement cost, so that it is the net present value which would be relevant for economic depreciation. In all preceding periods, it is the replacement value which is relevant, because the competitive product price generates a net present value that does not go below replacement cost.

What then are the effects of the decline in product prices in the last two periods on the capital theoretical profit neutrality of the machines? As the decline in prices is known ex ante, the firm must determine at which point in time the purchase of a new machine becomes unprofitable. This is already the case in period 5. Purchasing a new machine in period 5 would create a net present value $PV(6) = 10$ in period 6 and thereby cause an economic depreciation in period 5 of $PP(5)–PV(6) = 30$. This would result in a deficit of 10, which could not be balanced under the application of depreciation rule (2.6) in earlier periods, and thus would violate the capital theoretical profit neutrality. This scenario does not change, even if in earlier periods profits were made via monopolistic prices. In this case, economic depreciation would be determined by replacement cost, because replacement cost is lower than the net present value. Consequently, in period 5 only the old machine

2.1 Basic Principles of Determining Cost of Capital

will be utilised, which amortises in periods 4 and 5. Its net present value is 20 which equals the replacement cost of the machine bought in period 4.

2.1.3 Implementation

2.1.3.1 Conventional Depreciation Methods

In internal as well as external accounting, a multitude of depreciation methods are applied. They differ in particular with regard to the progress of depreciation over time, with linear, degressive, or progressive distribution, and the relative replacement cost or the utilisation of the asset.[4] The economic depreciation of a given asset can vary considerably within the planning horizon. While economic depreciation takes into account all relevant factors that are known at the beginning of the planning period (operating costs, wear and tear, prices on input markets, demand, etc.), this is not normally the case for the simple, time-dependent depreciation methods of historical costs and manufacturing costs. They do not characterise fluctuations in economic depreciation precisely, although such fluctuations do not occur unexpectedly in a stationary world. It is assumed that the depreciation of an asset will take place in constantly identical (linear depreciation), constantly decreasing (degressive depreciation), or constantly increasing (progressive depreciation) annual rates. Therefore, even in a world of stationary conditions conventional depreciation methods such as linear, degressive, or progressive depreciation can yield at best approximate results for determining decision-relevant costs.

2.1.3.2 Closed Versus Open Depreciation Schedules

Determining decision-relevant costs must be future-oriented, and thus cannot be done without expectations on the part of the decision-making firms. The problem of building expectations occurs in widely different model contexts. Of central importance is the question whether the expectations of economic agents with regard to future developments are "rational", that is, whether they are self-fulfilling. In this context, Muth characterises the rational expectation hypothesis as follows: "that expectations of firms (or, more generally, the subjective probability distribution of outcomes) tend to be distributed, ... about the prediction of the theory (or the 'objective' probability distributions of outcomes)" (Muth, 1961, p. 316). Whether rational expectations actually exist is of no concern at this point. Instead, what is important is the relevance of the assumption of rational expectations for determining decision-relevant costs.

Unforeseen changes of prices or technologies are usually excluded in the models for recording economic depreciation. Hotelling (1925, pp. 344f.) assumes that the value of a machine develops in a static or quasi-dynamic manner, and that at least the trend of this development of value can be reliably estimated at the beginning of the planning horizon. Turvey (1969, p. 298) excludes unforeseen changes of prices or technologies; otherwise the original depreciation schedule would not guarantee

---

[4] A numerical example for comparing the most relevant and well-established depreciation methods for price decreases can be found in Knieps et al. (2001, p. 763).

cost recovery. Other authors also exclude unexpected changes in replacement cost.[5] Wright (1968, p. 226) points out that depreciation schedules depend on the expectations at the beginning of the planning horizon. Thus, different expectations generate different depreciation schedules. Unforeseen changes of prices or technologies are disregarded. In the following it will become clear that depreciation in accordance with the deprival value principle makes possible adjustments to unforeseen changes of prices or technologies.

In order to understand the concept of economic depreciation, the models of closed depreciation schedules are indispensible. However, in dynamic economic sectors there is a necessity to react to changed economic conditions in the best possible way. "... it is obviously impossible to predict how machines, which have not even been invented as yet, will behave under economic conditions prevailing in the dim future." (Preinreich, 1940, p. 20). Technological progress and market changes that cannot be discerned ex ante at the beginning of the planning horizon cannot, by their very nature, be taken into consideration, and thus disregarding them should not ex post be described as an entrepreneurial failure. What remains is the necessity to adapt the depreciation schedules in a suitable way on the basis of the deprival value concept. This has to be done sequentially on the basis of current expectations. Sequential depreciation schedules are characterised by determining future depreciation values on the basis of current expectations, without considering past depreciation values, as the latter are no longer decision-relevant.

Price reductions for machines due to technological progress can be accounted for by closed depreciation schedules, insofar as rational expectations about future price developments can be generated at the beginning of the planning horizon. If, at the beginning of a period within the planning horizon, expectations regarding future machine prices change due to technological progress, the depreciation schedule has to be adapted accordingly.

In the following three cases are regarded. The starting point is the case study depicted in Table 2.1. However, it is now assumed that $PV(t) \geq RC(t)$ for all t, so that only replacement cost is relevant for depreciation. Because economic depreciation is the same for each unit of output, no matter if it is produced by the old or the new machine, in this special case—an economic life of 2 periods with identical production quantities in each period—it holds that $RC(t) = d(t)$. For simplicity, interest is assumed to be 0.

Case 1

Rational point expectation regarding machine prices at the beginning of period 1 (cf. Wright, 1968, p. 225)

The example regarded in Table 2.2 is also based on rational point expectation regarding future machine prices. Consequently, this results in a depreciation schedule which fulfils, from an ex ante perspective, the criterion of capital theoretical profit neutrality, that is, a schedule which finances the original investment

---

[5] Cf. Littlechild (1970, p. 330), Baumol (1971, pp. 650f.), Atkinson and Scott (1982, p. 20).

2.1 Basic Principles of Determining Cost of Capital

Table 2.2 Closed depreciation schedule

Year	1	2	3	4	5	6	7
$PP(t)$	100	70	50	40	40	40	40
$RC(t)$		40	30	20	20	20	20
$d(t)$	60	40	30	20	20	20	20

Table 2.3 Open depreciation schedule with one-time change of expectations

Jahr	1	2	3	4	5	6	7
$PP(t)$	100	70	50	20	20	20	20
$RC(t)$		30	40	10	10	10	10
$d(t)$	(60)	30	40	10	10	10	10

expenditure. The sum of the prices of the machines is 360. The sum of depreciations is also 360, with depreciation restricted to the new machine in the first period, to both machines equally $RC(t) + d(t) = 2d(t)$ in the following periods and the old machine being sold at replacement cost at the end of period 7.

Case 2
Change of expectations at the beginning of period 2, machine prices cut in half (from the originally expected 40–20) from period 4 onward (cf. Weiß, 2009, p. 73.).[6]

After the change of expectations at the beginning of period 2, an altered depreciation schedule results. The depreciation already transacted in period 1, however, has to be taken as given (cf. Table 2.3). The sum of purchase prices is 290 (the 7th machine is sold at the end of period 7 at replacement cost). The sum of depreciations is 280. The depreciation in period 1 could not be raised to 70 retroactively.

Case 3
First change of expectations at the beginning of period 2, machine prices cut in half from period 4 onward. Second change of expectations at the beginning of period 4, machine prices cut in half only from period 5 onward.

After the second change of expectations at the beginning of period 4, there will again be an altered depreciation schedule (cf. Table 2.4). The sum of purchase prices is 310 (the 7th machine is sold at the end of period 7 at replacement cost). The depreciations already undertaken to the amount of 200 cannot be altered retroactively. After two changes of expectations the remaining depreciations in the depreciation schedule are 120.

The sequential adaptation of depreciation schedules because of changed expectations regarding prices is necessary to comply with the fundamental principle of market reference. However, the adherence to capital theoretical profit neutrality

---

[6] The depreciations already undertaken before expectations changed are no longer relevant for the new depreciations schedule, and thus have been put in brackets.

Table 2.4 Open depreciation schedule with two-time changes of expectations

Year	1	2	3	4	5	6	7
$PP(t)$	100	70	50	40	20	20	20
$RC(t)$				30	10	10	10
$d(t)$	(60)	(30)	(40)	30	10	10	10

over the entire planning horizon from periods 1 to 7 is then no longer guaranteed. The sum of the relevant depreciations in each period is no longer equal to the sum of the purchasing prices of the machines. In case 2 there is a deficit of 10, in case 3 there is a surplus of 10.

If, due to technological progress and changed demand conditions, new decision-relevant information becomes available and reliable assumptions regarding future developments are not possible at the beginning of the planning horizon, that is, if there is an open event space, closed depreciation schedules fail to meet the criterion of market reference. Because of continuous changes that cannot be reliably estimated at the beginning of the planning horizon, it is necessary to periodically re-evaluate the relevant assets, taking into account the decision-relevant information on future developments available at each given point in time. Economic depreciation then can still be calculated as the difference between the value of the facility at the beginning of a given period and the value of the facility at the end of that period (cf. Knieps et al., 2001, pp. 768ff.).

2.1.3.3 Cost of Capital (Interest)

The cost of capital of a network provider should reflect the opportunity costs of the financial resources invested in assets within network industries. The basis is the concept of the Weighted Average Cost of Capital (WACC).[7]

$$\text{WACC} = r_E \cdot \frac{E}{E+D} + r_D \cdot (1-s) \cdot \frac{D}{E+D} \qquad (2.7)$$

where:
r_E interest rate on equity (rate of return demanded by shareholders)
r_D interest rate on debt (rate of return demanded by lenders of borrowed capital)
E market value of equity
D market value of debt
$E+D$ market value of total capital
s corporate tax rate

In this approach the weighted average cost of capital, based on the expected costs of debt and the expected costs of equity, is determined.

Modern corporate finance theory provides different approaches for determining interest rate on equity. The Capital Asset Pricing Model (CAPM) is the most

---

[7] This approach goes back to Modigliani and Miller (1958).

commonly used in corporate practice, due to its relatively transparent characterisation of the relevant risks (cf. Bruner, Eades, Harris, & Higgins, 1998).[8] The CAPM implies that a functioning capital market will produce expected rates of return, compensating investors for the non-diversifiable risk.

The aim of the CAPM is to derive a risk premium equivalent to the owner's risk. According to the CAPM, the equity provider's expected rate of return corresponds to the risk-free interest rate, plus the market risk premium (the difference between the expected rate of return of a fully diversified market portfolio and the risk-free interest rate) multiplied with the systematic risk (beta factor of the firm) (cf. e.g. Copeland, Koller, & Murrin, 2000, pp. 214ff.).

In order to determine the interest on the invested capital, both the equity and the debt have to be inserted into the WACC formula with their respective market values. An essential characteristic of forward-looking, decision-based cost of assets in a network industry is that it is based on the economic value, meaning the opportunity costs of the existing facility, which are reflected in the respective market values.

2.1.3.4 Disaggregated Determination of Cost of Capital

Cost of capital is incurred in the provision of network infrastructure capacities as well as in the provision of network services. Insofar as the suppliers of network services and the suppliers of network infrastructure capacities are separate firms, the cost of capital invested is also separated. Insofar as suppliers of network infrastructure also offer network services, the problem of how to differentiate between the costs of capital of different network levels arises.

Neither risk-free interest nor market risk premium are sector-specific. Systematic risk (beta) is naturally dependent on the specific entrepreneurial situation in the relevant sector, because it measures a firm's non-diversifiable risk as compared to a fully diversified market portfolio. Corporate finance theory demands a forward-looking beta which measures the expected future risks of an investment in a firm, relatively to the market portfolio.

In the context of cost-based regulations in liberalised network industries an intense parameter debate on the determination of decision-relevant risk parameters between regulated firms and regulatory agencies can be observed (cf. Knieps, 2003). A fundamental distinction has to be made between the risk on the markets for network services and the risk on the markets for network infrastructure capacities. In case of vertically integrated network companies the starting point is to determine the specific risk of the respective business division. This goes hand in hand with the problem of determining the specific costs of equity for each business division. But the systematic risk (beta factors of the firms) also strongly varies

---

[8] In the context of this paper it does not seem practical to examine the theoretical foundation of the different approaches of modern finance theory to determining the interest rate on equity, and their respective strengths and weaknesses.

between different network infrastructures, resulting in different WACCs.[9] In particular, in dynamic subparts of network industries the systematic risk is higher compared to rather stationary subparts.

2.2 Decision-Relevant Cost Allocation

Depending on the underlying decision situation, different cost concepts are relevant. As network companies typically are multi-product firms, cost concepts have to be applied in this context. Depending on the existing decision situation, the following micro-economically well-founded cost concepts could be relevant:

- short-run marginal costs in order to determine usage-dependent service rates;
- long-run marginal costs in order to take into account the user cost of capital that results from providing the product;
- long-run incremental costs in order to determine, whether additional network services or network infrastructures should be supplied;
- long-run avoidable costs in order to determine, whether certain network services or network infrastructures should no longer be supplied;
- stand-alone costs of products or product groups in order to determine whether the building of alternative sub-networks is cost-covering;
- total costs in order to determine the viability of the active network providers.

2.2.1 Short-Run Versus Long-Run Marginal Costs

Long-run marginal costs are different from short-run marginal costs, in that for the former, the (periodic) user cost of capital of capacity provision also has to be taken into account. While capacity is fixed over a given period of time, short-run fluctuations in usage within the capacity limit are possible. It is necessary to differentiate between a scenario where capacity can be adapted at the beginning of each period, so that at any given point in time the optimum combination of fixed and variable costs can be chosen, and a scenario where such an adaptation is impossible, or only conditionally possible. In the latter case, the cost structure of the industry is path dependent, because in each period it is dependent on the history of investments made, technological developments, and the development of relative factor prices (cf. Turvey, 1969, pp. 285f.). Even if the combination of fixed and variable costs is variable over time, the provision of capacity represents fixed costs that do not fluctuate with the degree of utilisation of the facility. While long-run marginal costs include the fixed costs of capacity provision, short-run marginal costs are a function of the usage intensity (up to the capacity limit).

---

[9] For the case of different regulated network industries in Great Britain see Knieps (2003, p. 1001, Table 1).

2.2.2 Long-Run Incremental Costs

In the wake of the comprehensive opening of networks and the resultant problem of network access the concept of incremental costs has become more relevant in the last decades; the roots of this approach, however, go back as far as the nineteenth century (cf. Alexander, 1887). In the meantime, the concept has been refined with the help of game theoretical methods and developed further for application to product groups (cf. Faulhaber, 1975; Knieps, 1987).

Long-run incremental costs: In a multi-product case with $N = \{1, \ldots, n\}$ products, the long-run incremental costs of a product $i \in N$ when producing an output vector $y = (y_1, \ldots, y_n)$ are defined as:

$$\overline{C}(y_i) = C(y) - C(y_{N-i}), \quad \text{where} \quad y_{N-i} = (y_1, \ldots, y_{i-1}, 0, y_{i+1}, \ldots, y_n) \quad (2.8)$$

Consequently, these are the additional costs resulting from one additional product i being manufactured, provided that all other products are being manufactured in any case. The long-run average incremental costs of a product i are thus defined as (cf. Baumol, Panzar, & Willig, 1982, p. 67):

$$A\overline{C}(y_i) = \frac{\overline{C}(y_i)}{y_i} \quad (2.9)$$

The long-run incremental costs of a product group $S \subset N$ when producing an output vector $y = (y_1, \ldots, y_n)$ are defined as

$$\overline{C}(y_S) = C(y) - C(y_{N-S}), \quad (2.10)$$

where y_S denotes the vector for which $y_i > 0$ for $i \in S$ and $y_i = 0$ for $i \notin S$

The long-run incremental costs of a product group denote the additional costs resulting from one additional product group being supplied, if all other product groups are being supplied in any case. Incremental costs can be fixed or variable.

When applying the concept of incremental costs, the definition of the relevant additional product, or the relevant additional product group, is of crucial importance. The concept of incremental costs can only be applied to outputs. The definition of additional products is an entrepreneurial decision. It is in particular dependent on market circumstances. No fictitious products, but currently supplied or planned products are taken into account. In the special case of a one-product firm the long-run incremental costs are identical to the long-run marginal costs.

2.2.3 Long-Run Incremental Costs Versus Long-Run Avoidable Costs

The differentiation between fixed costs and irreversible costs is of fundamental importance (cf. Baumol, 1996, pp. 57f.). Fixed costs can be either irreversible or reversible. While reversible fixed costs can be applied flexibly in different markets, irreversible costs, once spent on a particular market, cannot be utilised elsewhere. An airplane represents fixed costs, which can be used on different markets und consequently is not irreversible. An airport, on the other hand, represents irreversible costs. If there are irreversible costs, it is necessary to distinguish between long-run incremental costs and long-run avoidable costs. For the decision to enter the market, irreversible costs are also relevant, whereas for the decision to leave the market, only avoidable costs are relevant, if the facility is not going to be used otherwise.

As long as a market is not abandoned (e.g. by closing down an airport), facilities with sunk (irreversible) costs also have market-oriented opportunity costs of utilisation and thus are also a component of the decision-relevant incremental costs. At the same time, there is path dependency regarding decisions on network utilisation. Thus it can, for instance, be less costly to expand an airport when demand increases, instead of replacing it with a brand-new facility. The economic value of facilities with sunk costs can be very high, depending on the discounted future revenues that can be achieved.

2.2.4 The Traditional Concept of Overhead Costs

The concept of overhead costs has a long tradition. In his classical book, Clark defines the term overhead costs as follows:

> What are 'overhead costs'? ... They refer to costs that cannot be traced home and attributed to particular units of business in the same direct and obvious way in which, for example, leather can be traced to the shoes that are made from it (Clark, 1923, p. 1).

In this context he emphasises that even under competition there are various differentiation options for covering overhead costs (cf. Clark, 1923, p. 32). The basic principle is to systematically utilise the different degrees of willingness to pay of different demand groups in the firm's pricing strategies. Consequently, the mark-ups required for covering overhead costs have to be chosen depending on the price elasticity of demand. The lower the elasticities, the higher the mark-ups should be set. The price setting follows the principle "charging what the traffic will bear" (Clark, 1923, p. 281). Price differentiation according to the elasticities of different user groups is also necessary under competition, in order to guarantee the covering of overhead costs.

The starting point of all cost allocation rules developed in this context is the differentiation between the costs resulting from offering single additional services (incremental costs) on the one hand, and the residual costs (overhead costs) on the other.

2.2 Decision-Relevant Cost Allocation

Overhead costs are defined as follows:

$$OC = C(y_N) - \sum_{i \in N} \overline{C}(y_i) = C(y_N) - \sum_{i \in N} (C(y_N) - C(y_{N-i})) \quad (2.11)$$

For simplicity, in the following a discrete representation of the costs is chosen. Let $C(N)$ denote the costs of providing all $N = \{1, \ldots, n\}$ services through one single firm. Then the representation of overhead costs is:

$$OC = C(N) - \sum_{i \in N} \overline{C}(i) = C(N) - \sum_{i \in N} (C(N) - C(N-i)) \quad (2.12)$$

Thus costs that are exclusively incurred in producing a subset (coalition) of services are still, according to the traditional understanding of the term, overhead costs. Consequently, the problem of overhead costs is finding an unambiguous rule for dividing up all costs that cannot be charged to an individual service as incremental costs.

2.2.5 Product-Group Specific Common Costs and Firm-Specific Common Costs

Product-group specific common costs are cost advantages resulting from producing a group of services (components) together within a single firm. Although these costs can be allocated to a specific product group, they cannot be causally allocated to an individual product either directly or indirectly. In contrast, firm-specific common costs cannot be causally allocated either to individual products or to product groups either directly or indirectly, as long as more than one product group is produced. If only one product group is produced within a firm $C(S)$ denotes the costs of providing a group of services through a separate firm. The stand-alone costs of this product group thus consist of the product-specific incremental costs of each product, the product-group specific common costs and the firm-specific common costs.

Consider a case with three products: $C(i)$, $i = 1, 2, 3$ are the costs for producing product i separately; $C(1,2)$, $C(1,3)$, $C(2,3)$ are the costs for producing one individual product group and $C(1,2,3)$ are the costs of a combined production of the 3 products. Joint production represents the most extreme case of economies of scope in the combined production of goods in one firm. If

$$\begin{aligned} C(1) &= C(2) = C(3) = C(1,2) = C(1,3) = C(2,3) \\ &= C(1,2,3), \text{ it is a case of joint production.} \end{aligned} \quad (2.13)$$

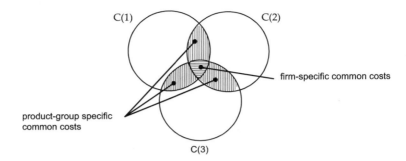

Fig. 2.2 Product-group specific common costs in the three-product case

Economies of scope are a much more common phenomenon than joint production.[10] They occur whenever it is more cost-efficient to produce a (sub-)set of products together in one firm. The costs of combined production are then lower than the costs of separate production. Although some of the production facilities are used in combination, the different products are not produced in fixed proportions, irrespective of the dimension of the combined production facility. In particular, the costs of producing one product separately (stand-alone costs) are lower than the costs of simultaneously producing an additional product, because the incremental costs of the latter product can be saved. This is already the crucial difference between economies of scope and joint production. Beside the product-group specific common costs, there are also the firm-specific common costs, irrespective of how many products the firm manufactures.[11]

Fig. 2.2 refers to the three-product case (cf. Knieps, 2008, p. 38). The firm-specific common costs $C(1) \cap C(2) \cap C(3)$ are horizontally hatched and the product-group specific common costs

$\{C(1) \cap C(2)\} - \{C(1) \cap C(2) \cap C(3)\}$, $\{C(1) \cap C(3)\} - \{C(1) \cap C(2) \cap C(3)\}$, and $\{C(2) \cap C(3)\} - \{C(1) \cap C(2) \cap C(3)\}$ are vertically hatched.

Combined, the vertically hatched product-group specific common costs and the horizontally hatched firm-specific common costs represent the overhead costs.

If there are economies of scope, a cost allocation problem will inevitably develop, because neither marginal costs nor incremental costs (including product-group specific fixed costs) are sufficient to cover total costs and thus guarantee the viability of the firm. The mark-ups on the incremental costs of a product for covering product-group specific common costs and firm-specific common costs require the involvement of the demand side, and thus the use of price differentiation strategies (cf. Chap. 4). Although it may be difficult to ascertain the customers'

---

[10] Under competition economies of scope are sufficient for the existence of multi-product firms (cf. Panzar & Willig, 1981).

[11] It is, however, important to remember that overhead costs constitute a dynamic phenomenon (cf. Clark, 1923, pp. 478f.) and that even firm-specific common costs can be lowered (at least in the longer run) by reducing the range of products.

price elasticities exactly (information problem), this should not lead to neglecting the demand side completely and employ arbitrary common cost allocation schemes (cf. Baumol, Koehn, & Willig, 1987). Instead, it is important to develop suitable criteria, on the basis of which substitution options of different products or product groups can be estimated. Such criteria might include components taking into account the time of day or the season, but also information on suitable substitution options. Although information on individual willingness-to-pay may not be available, estimates of the average valuation of the product among all individuals in a given demand group are definitely possible.

Decision-based cost allocation, however, requires a differentiated procedure. Thus the revenues from each product have to cover not only its product-group specific incremental costs but also contribute to the recovery of the product-group specific common costs and the firm-specific common costs. If the individual products of a product group together are unable to bear their product-group specific common costs, they will no longer be produced. Likewise, all products together must cover the firm-specific common costs in order to ensure the viability of the firm.

2.3 Cost Strategies in Networks

2.3.1 Network Evolution

Decision-relevant costing requires that costs are determined in a forward-looking manner and on the basis of an efficient network. It is based on concrete network providers, either building their own networks, or operating on the basis of an existing network. In the case of an existing network, its path dependency has to be taken into account. The path dependency of investments describes the fact that a firm will make new investments, and decide on expanding existing networks or building new ones, depending on the entirety of the investments it has already made in the past.

The search for the optimal network structure is a complex entrepreneurial challenge faced by every network provider, as different decision parameters have to be taken into account simultaneously, such as strategies for product differentiation, network capacity and network quality. This goes together with decisions on efficient network expansion, taking path dependency into account. Unfulfilled prognoses regarding network utilisation represent entrepreneurial risks and should not automatically be equated with inefficiencies.

2.3.2 Strategies for Building a Network

It is necessary to differentiate between the decision-relevant costs of network infrastructure provision and network services provision. No matter if network infrastructures are newly built or if existing network infrastructures are expanded, these are entrepreneurial decisions.

- Network Development
  Three fundamental types of network development can be differentiated:

  1. A completely new network is planned and built from scratch.
  2. Based on an existing network topology, existing networks are completely rebuilt and replaced by new facilities.
  3. Existing networks are modernised by replacing individual parts of the network with more efficient ones. Parts of the existing infrastructure remain in use unmodified.

- Specialised versus Multipurpose Infrastructures
  It is important to differentiate between specialised network infrastructures, which are built for only one type of utilisation (water networks, electricity networks, natural gas networks) and multipurpose network infrastructures, which allow for several types of utilisation (interactive cable-TV networks, rail tracks for freight transportation and passenger transportation). However, it is also possible to provide different network services on specialised network infrastructures.

- The Significance of Path Dependency
  One has to differentiate between the significance of path dependency on the network infrastructure level on the one hand, and on the network services level on the other. Alternative investment strategies (building, expansion, close-down, etc.) take the path dependency of network infrastructures into account. On the network services level, path dependency is of lesser significance, although timetables for providing network services cannot be changed in the short-term either.

  The question whether certain network infrastructure capacities or network services should be discontinued or additionally supplied can only be answered on the basis of the long-run incremental costs or the long-run avoidable costs of a network provider. Only the costs determined on the basis of an actually existing network are relevant, because only these costs reflect the economic evaluation of resources which would be saved, if these additional services were not or no longer provided. The long-run incremental costs have to be determined on the basis of a firm's real cost data. As long as the additional costs of the expansion of an existing network are lower than the costs of building a new network, the further development of the existing network is profitable. Therefore the development of networks is path dependent. In particular, the path dependency of a network provider's investment decisions must not be neglected.

  However, weighing the incremental costs on the basis of existing network topologies against the costs of building a new network is not at all the same as clinging to traditional network technology. Building parts of networks completely new can be profitable under certain conditions. Thus a differentiated analysis of the necessary expansion strategies is of crucial importance.

- The Option Value of Delayed Investments
  There are various reasons why investments in network industries can become obsolete sooner than expected and the facilities have to be shut down. Among those reasons are unexpected drops in demand, as well as the emergence of new, competing technologies. From the investor's ex ante perspective there is, however, also the possibility that the investment will yield an above average profit. Unexpected changes in demand and technology constitute entrepreneurial risks which have to be taken into account in the context of a market-based interest rate on the invested capital via appropriate risk mark-ups.

  If there is uncertainty of demand in combination with irreversible cost, there will be an asymmetry between an investment at the present moment and future investments. Today's investment cannot be used elsewhere, whereas the capital not yet invested can be utilised with complete flexibility, depending on future market developments. The value of these future options due to delayed investment activity is not represented by the net present value. To the degree that the option value of postponing investments is significant, it should be taken into account when calculating the interest on equity and consequently also when considering long-run incremental costs. The value of the real option to postpone investments corresponds to the opportunity costs of investing today. This result can be derived from the application of modern theories of investment under uncertainty (cf. Dixit & Pindyck, 1994; MacDonald & Siegel, 1986; Trigeorgis, 1999).

2.3.3 Decision-Relevant Costs of New Networks

In order to decide whether the building of alternative networks is profitable, it is necessary to determine the stand-alone costs of hypothetical networks. Analytical cost models can be very helpful in network providers' decision processes.

Analytical cost models are derived from process analysis models which were developed in the 1950s for creating hypothetical cost data (as opposed to real cost data) for isolated facilities, such as refineries, electricity generation, etc. (cf. Manne, 1958). The basic idea underlying these models is that in a first step a production function is modelled on the basis of engineering data and then, based on this, cost functions assuming optimisation behaviour (duality theory) with given input prices are derived. The cost data thus derived are not real data from cost accounting, but simulated data; their informational value is crucially dependent on the quality and completeness of the underlying process model (cf. Griffin, 1977, p. 125).

Analytical cost models were introduced to network economics in the context of the debate on the natural monopoly characteristics of telecommunications networks. Although over many years process analyses did not receive much attention, since the beginning of the 1990s they have been applied to telecommunications networks. The objective of these analytical cost models is to examine the costs of alternative hypothetical network structures. It should be possible to choose both the combination and the location of switching facilities in such a way that the production costs are minimised for different demand levels (cf. Gabel & Kennet, 1994, p. 386). Due to the non-linearity of the objective function, the search for solutions is undertaken via simulation processes. As it is necessary to limit the simulation to an "appropriate" number of possible solutions, this is not a global approach, but a bounded rationality approach with considerable degrees of freedom in the search for "plausible" solutions. In order to determine the stand-alone costs of novel network parts the application of analytical cost models can be very useful. This, however, presupposes that the firm itself makes the assumptions on the limitation of the degrees of freedom regarding the number and location of switching facilities.

For determining decision-relevant costs in existing networks analytical cost models are not suitable. Even if, in the context of the optimisation model, the existing network topology is taken as the basis, important real cost elements indispensable for economically well-founded costing are neglected. The effects of disregarding the path dependency of investment planning are particularly problematic.

2.3.4 Long-Run Incremental Costs of Novel Network Services

It is necessary to differentiate between the long-run incremental costs of providing novel network services and the long-run incremental costs this causes on the level of network infrastructure capacity. Thus serving flight routes with higher quality airplanes causes higher incremental costs, which have to be covered via the prices of the plane tickets. The situation is different with regard to the question whether the network infrastructure used for the provision of higher quality network services has to be exclusively financed by the users of these higher quality network services. The answer to this question depends crucially on the point in time (before or after the decision about a particular network extension) und the assumptions on the future development of the network infrastructure. An economically well-founded valuation of incremental costs is dependent on the entrepreneurial strategy for network evolution in the context of the provision of new network services.

The controversy over the economically well-founded valuation of incremental cost has its roots in different perceptions of network evolution in the context of the provision of new telecommunications services. Kahn and Shew (1987) consider the essential problem of network expansion already solved and, in particular, assume that it is not efficient to build a completely new network in order to satisfy additional demand (cf. Taylor, 1993, pp. 32f.). Gabel and Kennet (1993, 1994), on the contrary, hold the view that specialised networks as well as multipurpose networks will be completely optimised, with even network topology (for example the location and number of switching equipment) considered endogenous. As long as the gradual expansion of telecommunications networks is efficient, these constitute multipurpose networks as described by Kahn and Shew (1987), as well as

Taylor (1993). The situation is different, however, if it is not an expansion of networks, but rather the provision of network infrastructure variety which constitutes the economically efficient solution (depending on the demand for different services, the distribution of customers over space, etc.). Examples for this are interactive cable TV networks and broadband telecommunications networks. The approach by Kahn and Shew (1987, p. 231), interpreting the decision for an integrated telecommunications network and the resultant quality of integrated network access as a collective consumer decision, falls short in this context. For the costing of newly to be built parallel networks the optimisation of different networks can play an important role (cf. Gabel & Kennet, 1993, 1994).

2.4 Questions

2-1: User Cost of Capital
Explain the principle of capital theoretical profit neutrality, as well as the principle of market reference.

2-2: Deprival Value
How is the value of a facility or a machine calculated at a given point in time according to the deprival value concept?

2-3: Economic Depreciation
Explain the difference between closed and open depreciation schedules.

2-4: Path Dependency and Network Evolution
Explain the term path dependency and its relevance for the building and further development of networks.

2-5: Long-Run Incremental Costs of Novel Network Services
Explain the problems of calculating the long-run incremental costs of novel network services.

References

Alexander, E. P. (1887). *Railway practice*. New York: G.P. Putnam's Sons.
Atkinson, A. A., & Scott, W. R. (1982). Current cost depreciation: A programming perspective. *Journal of Business Finance and Accounting, 9*(1), 19–42.
Baumol, W. J. (1971). Optimal depreciation policy: Pricing the products of durable assets. *Bell Journal of Economics, 2*(2), S.638–S.656.
Baumol, W. J. (1996). Predation and the logic of the average variable cost test. *Journal of Law and Economics, 39*, 49–72.
Baumol, W. J., Koehn, M. F., & Willig, R. D. (1987). How arbitrary is "arbitrary" or, toward the deserved demise of full cost allocation. *Public Utilities Fortnightly, 120*(5), 16–21.

Baumol, W. J., Panzar, J. C., & Willig, R. D. (1982). *Contestable markets and the theory of industry structure*. San Diego: Harcourt Brace Jovanovich.

Bell, P. W., & Peasnell, K. (1997). Another look at the deprival value approach to depreciation. In T. E. Cooke & C. W. Nobes (Eds.), *The development of accounting in an international context* (pp. 122–148). London: Routledge.

Bruner, R. F., Eades, M., Harris, R. S., & Higgins, R. C. (1998). Best practices in estimating the cost of capital: Survey and synthesis. *Financial Practice and Education, 8*(1), 13–28.

Clark, J. M. (1923). *Studies in the economics of overhead costs*. Chicago: The University of Chicago Press.

Copeland, T., Koller, T., & Murrin, J. (2000). *Valuation – Measuring and managing the value of companies* (3rd ed.). New York: Wiley.

Dixit, A., & Pindyck, P. (1994). *Investment under uncertainty*. Princeton: Princeton University Press.

Enke, S. (1962). Production functions and capital depreciation. *The Journal of Political Economy, 70*(4), 368–379.

Faulhaber, G. R. (1975). Cross-subsidization: Pricing in public enterprises. *The American Economic Review, 65*(5), 966–977.

Gabel, D., & Kennet, D. M. (1993). Pricing of telecommunications services. *Review of Industrial Organization, 8*, 1–14.

Gabel, D., & Kennet, D. M. (1994). Economics of scope in the local telephone exchange market. *Journal of Regulatory Economics, 6*, 381–398.

Griffin, J. M. (1977). Long-run production modeling with pseudo-data: electric power generation. *Bell Journal of Economics, 8*, 112–127.

Hotelling, H. (1925). A general mathematical theory of depreciation. *Journal of the American Statistical Association, 20*, 340–353.

Kahn, A. E., & Shew, W. B. (1987). Current issues in telecommunications regulation: Pricing. *Yale Journal on Regulation, 4*(2), 191–256.

Knieps, G. (1987). Zur Problematik der internen Subventionierung in öffentlichen Unternehmen. *Finanzarchiv, N.F. 45*, 268–283.

Knieps, G. (2003). Entscheidungsorientierte Ermittlung der Kapitalkosten in liberalisierten Netzindustrien. *Zeitschrift für Betriebswirtschaft, 73*, 989–1006.

Knieps, G. (2008). *Wettbewerbsökonomie – Regulierungstheorie, Industrieökonomie, Wettbewerbspolitik* (Springer-Lehrbuch, Vol. 3). Berlin: Springer.

Knieps, G., Küpper, H.-U., & Langen, R. (2001). Abschreibungen bei fallenden Wiederbeschaffungspreisen in stationären und nicht stationären Märkten. *Schmalenbachs Zeitschrift für betriebswirtschaftliche Forschung (zfbf), 53*, 759–776.

Littlechild, S. C. (1970). Marginal-cost pricing with joint costs. *The Economic Journal, 80*(318), 323–335.

MacDonald, R., & Siegel, D. (1986). The value of waiting to invest. *Quarterly Journal of Economics, 1001*, 707–728.

Manne, A. (1958). A linear programming model of the U.S. petroleum refining industry. *Econometrica, 26*, 67–106.

Modigliani, F., & Miller, M. H. (1958). The cost of capital, corporation finance and the theory of investment. *American Economic Review, 48*, 261–297.

Muth, J. F. (1961). Rational expectations and the theory of price movements. *Econometrica, 29*(3), 315–335.

Panzar, J., & Willig, R. D. (1981). Economies of scope. *American Economic Review, 71*(2), 268–272.

Preinreich, G. A. (1940). The economic life of industrial equipment. *Econometrica, 8*(1), 12–44.

Solomons, D. (1966). Economic and accounting concepts of cost and value. In M. Backer (Ed.), *Modern accounting theory* (pp. 117–140). Engelwood, Cliffs: Prentice-Hall.

Taylor, W. E. (1993). Efficient pricing of telecommunications services: The state of the debate. *Review of Industrial Organization, 8*, 21–37.

References

Trigeorgis, L. (1999). Real options: An overview. In J. Alleman & E. Noam (Eds.), *The new investment theory of real options and its implication for telecommunications economics* (pp. 3–34). Boston: Kluwer.

Turvey, R. (1969). Marginal cost. *The Economic Journal, 79*(314), 282–299.

Vickrey, W. (1985). The fallacy of using long-run cost for peak-load pricing. *Quarterly Journal of Economics, 100*, 1331–1334.

Weiß, H.-J. (2009). *Entscheidungsorientiertes Costing in liberalisierten Netzindustrien* (Freiburger Studien zur Netzökonomie, Vol. 16). Baden-Baden: Nomos.

Wright, F. K. (1968). Measuring asset services: A linear programming approach. *Journal of Accounting Research, 6*(2), 222–236.

Congestion Externalities 3

3.1 Local (Path-Based) Externalities

The phenomenon of congestion can be observed on the markets for network services (level 1) as well as on the markets for infrastructure capacities (levels 2 and 3).

3.1.1 Congestion Externalities and Congestion Fees

Congestion externalities can be involved in the use of any transportation infrastructures, such as airports, railway tracks, ports and roads. The field that has been studied most intensively in the literature of transport economics is road congestion externalities. Therefore, in the following we will first examine the congestion problem on motorways.

Traffic participants tend to ignore the burden imposed on other traffic participants by an additional vehicle at a particular moment, such as for example longer clearance times, longer delays, and longer driving times. These are negative physical externalities, which—in contrast to monetary externalities—are not internalised via market prices.

In the following the marginal congestion costs of one trip on a transportation infrastructure will be specified in more detail (cf. Dewees, 1979). Centrally important for this is traffic flow Q, that is, the number of vehicles using a given section of motorway during a specific time period.[1] Maximum traffic flow $\overline{Q}$ denotes the capacity of the infrastructure. In order to simplify the scenario we will assume that traffic on the infrastructure under consideration is homogenous, so that driving time and valuation of time are identical for all vehicles. We will examine a traffic flow of Q vehicles per hour, requiring a time period of T to pass through a 1-km road

---

[1] In order to simplify notation time index t is neglected in this section.

section at speed v. Speed v is dependent on traffic flow Q (with $\frac{\partial v}{\partial Q} < 0$). One vehicle has driving costs of $\frac{c}{v(Q)}$ per kilometre, where c denotes the time costs per vehicle hour. The total time costs for a traffic flow of Q vehicles per hour to pass through this 1-km road section is thus $\frac{c \cdot Q}{v(Q)}$. If one additional vehicle enters the infrastructure, total time costs increase by:

$$\frac{\partial}{\partial Q}\left(\frac{c \cdot Q}{v(Q)}\right) = \frac{c \cdot v - c \cdot Q \frac{\partial v}{\partial Q}}{v^2} = \frac{c}{v} - \frac{c \cdot Q}{v^2} \cdot \frac{\partial v}{\partial Q} \qquad (3.1)$$

For the average speed v on the relevant road section (1 km) it holds true that $v = \frac{1}{T}$ and thus:

$$\frac{\partial}{\partial Q}\left(\frac{c \cdot Q}{\frac{1}{T(Q)}}\right) = cT(Q) + Qc\frac{\partial T}{\partial Q} \qquad (3.2)$$

The first term cT denotes the time costs that must be borne by the additional vehicle itself. The marginal congestion costs that the additional vehicle imposes on all other vehicles in traffic flow Q are characterised by the second term $Qc\frac{\partial T}{\partial Q}$. The longer travel time imposed by one additional vehicle on all other vehicles is thus given by $Qc\frac{\partial T}{\partial Q}$.

In addition to the time costs $cT(Q)$, the variable costs $k(Q)$ of one vehicle utilising the road also consist of the operating costs of the vehicle (fuel, maintenance, etc.). To this are added the utilisation-dependent maintenance costs of the underlying infrastructure. The operating costs of the vehicle may also depend on traffic flow. The congestion costs that one additional vehicle imposes on all other vehicles are determined analogous to pure time cost calculation. The derivation of total costs $k(Q) \cdot Q$ results in the social marginal costs MC_s of one additional trip:

$$MC_s = \frac{\partial}{\partial Q}(k(Q) \cdot Q) = k(Q) + \frac{\partial k}{\partial Q} \cdot Q \qquad (3.3)$$

where $k(Q)$ denotes the variable costs for the additional vehicle and $\frac{\partial k}{\partial Q} \cdot Q$ denotes the congestion costs for all other vehicles in traffic flow Q.

3.1.2 Optimal Congestion Fees

One possible measure for achieving optimal traffic flow would be to levy a (time-dependent) congestion fee equal to the congestion costs incurred by all other vehicles as a result of the one extra trip. Only then can it be ensured that each vehicle bears the social marginal costs of its road utilisation.

3.1 Local (Path-Based) Externalities

The fundamental idea of interpreting congestion as an externality which is imposed by, for instance, one extra vehicle on all other vehicles on a section of motorway, immediately leads to the concept of externality costs, which has been known since Pigou (1920). While Pigou, who understood transportation infrastructures as belonging to the public sector, suggested a usage-dependent special tax solution for the internalisation of congestion externalities, Knight (1924, pp. 584–590), assuming private transportation infrastructures, argued that under competition the provider would have an inherent incentive to limit usage to the optimal level by raising a congestion fee equal to congestion costs.[2] Thus it is shown that under competition the profit-maximising congestion fee is equal to the socially optimal congestion fee.

Under competition, the profit-maximising price must be equal to social marginal costs. Let $P(Q)$ denote the inverse demand function for trips on a particular road section. Then it holds that

$$\max_Q \pi = P(Q) \cdot Q - Q \cdot k(Q) \tag{3.4}$$

$$P^w = \frac{\partial(k(Q) \cdot Q)}{\partial Q} = k(Q) + \frac{\partial k}{\partial Q} \cdot Q = MC_s \tag{3.5}$$

Congestion fee under competition then is:

$$\tau^w = P^w - k(Q) = \frac{\partial k}{\partial Q} \cdot Q \tag{3.6}$$

The profit-maximising gross price does not only include the variable costs of the marginal vehicle, but also the externality costs which result from the higher costs imposed on all vehicles in the traffic flow.

Maximising social welfare S results in[3]:

$$\max_Q S = \int_0^Q P(\widetilde{Q}) d\widetilde{Q} - Qk(Q) \tag{3.7}$$

and thus:

$$\frac{\partial S}{\partial Q} = P^o - \left[k(Q) + \frac{\partial k}{\partial Q} \cdot Q\right] = 0 \tag{3.8}$$

---

[2] An overview of the history of road pricing can be found, e.g., in Button (2004, pp. 8–17), Mohring (1999, pp. 193–198).

[3] It is assumed that the income effects associated with the inverse demand function are negligible. The social welfare (social net utility) of trips on a given road section is characterised as the sum of consumer and producer surplus.

Fig. 3.1 Socially optimal congestion fees

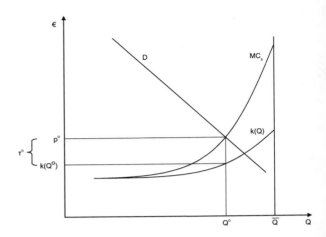

The socially optimal congestion fee τ^o thus is:

$$\tau^o = P^o - k(Q) = \frac{\partial k}{\partial Q} \cdot Q \tag{3.9}$$

Consequently, under competition the provider of a privately owned road will raise a congestion fee τ^w that is equal to the socially optimal congestion fee. The socially optimal congestion fee is illustrated in Fig. 3.1.

The socially optimal congestion fee reflects congestions costs. In addition, drivers must carry the private variable costs including the utilisation-dependent maintenance costs of the road infrastructure.[4]

3.1.2.1 Capacity Scarcity and Hyper-Congestion

Up to capacity limit $\overline{Q}$ an increase in the number of vehicles on a road section leads to higher traffic flow. The reduction of speed due to higher traffic density θ (number of vehicles on the road section) has less influence on traffic flow than the increase in traffic flow due to a further increase in the number of vehicles. Under these conditions it is sufficient to examine the effects on congestion costs caused by a change in traffic flow (due to the increase in traffic density), and thus determine the socially optimal congestion fees for achieving optimal traffic flow. In centrally coordinated network infrastructures (e.g. railway tracks, airports) traffic density beyond capacity limit is not permitted to occur.

The situation is different for road traffic, where, depending on time of day, time of year, or geographical location, capacity limits can be exceeded

---

[4] Maintenance costs which depend on traffic volume using the road are to be distinguished from maintenance costs which are variable with road size. Maintenance costs variable with road size are part of the investment problem (cf. Keeler & Small, 1977, p. 3).

3.1 Local (Path-Based) Externalities

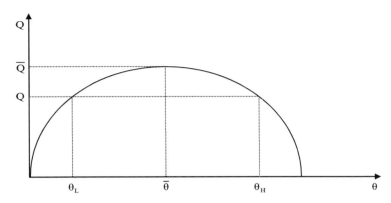

Fig. 3.2 Traffic density and traffic flow

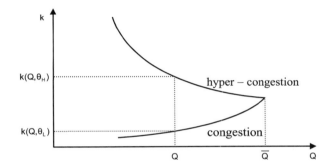

Fig. 3.3 Congestion and hyper-congestion

(cf. Figs. 3.2 and 3.3), so that a hyper-congestion scenario occurs. If traffic density increases beyond capacity limit ($\theta > \bar{\theta}$), traffic flow decreases. In the marginal case of maximum traffic density, there is a row of cars not moving forward at all, that is, a traffic flow of zero, so that traffic stops completely.

As soon as capacity limits are exceeded, any traffic flow $Q < \bar{Q}$ can be achieved, either for high traffic density θ_H (with low speed) or for low traffic density θ_L (with high speed).[5] From this follows that for traffic densities $\theta > \bar{\theta}$ cost inefficiencies will occur (cf. Button, 2004, p. 6).

Because any traffic flow $Q < \bar{Q}$ can be assigned values $k(Q, \theta_L)$ and $k(Q, \theta_H)$, this is not a cost function but a cost correspondence (set-valued cost relation) with one efficient and one inefficient branch. In the context of transport economics the topic of hyper-congestion gains increasing attention; in the context of the analysis of congestion models, however, it is still widely disregarded (cf. Button, 2004,

---

[5] The relation between traffic density and traffic flow has already been examined in Büttler (1982, p. 186) and Mohring (1999, pp. 182ff.).

Fig. 3.4 Capacity constraints

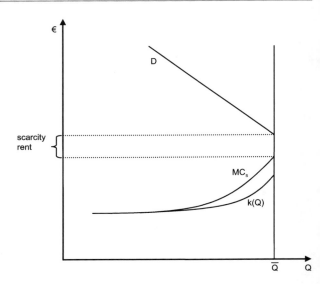

pp. 4–7; Mohring, 1999, pp. 182–187). A fundamental distinction has to be made between over-usage of a transportation infrastructure in the inefficient range beyond capacity limits (hyper-congestion) and over-usage in the efficient range.

3.1.2.2 Capacity Scarcity and Scarcity Rents

In the following the hyper-congestion scenario will be excluded, so that for every infrastructure a degree of utilisation in keeping with capacity limit $\overline{Q}$ will be assumed. If, with optimal congestion fees being levied, at certain times the capacity limit is reached and an excess demand for network capacity remains, it becomes necessary to raise a scarcity rent. The optimal solution is to raise a (time-dependent) scarcity price, which balances the supply, limited by the capacity of the network infrastructure, against the demand of the infrastructure users (cf. Fig. 3.4).

The results can be directly transferred to other network sectors. Based on the demand for network services (level 1), congestion fees, or scarcity prices for access to a network infrastructure, respectively, can be derived. Examples for this are access tariffs to railway tracks or airports (cf. e.g. Knieps, 2006).

3.1.3 Socially Optimal Congestion Fees and Investment Decisions

3.1.3.1 Congestion Fees for a Given Transportation Infrastructure

In the previous sections the basic principle of optimal congestion fees has been demonstrated, without explicitly taking into account fluctuations in capacity utilisation over time. In the following, the demand for trips in different time periods is examined. The optimal congestion fees for a given transportation infrastructure in

3.1 Local (Path-Based) Externalities

the case of a homogenous user group, depending on the particular point in time t, are derived analytically as follows[6]:

Let $P_t = P_t(Q_t)$ denote the inverse demand function for vehicle trips in period t on a given section of motorway. For the sake of simplification we will assume that $P_t(Q_t)$ is independent of the individual periods. Let Q_t denote traffic flow, that is, the number of vehicles passing through the road section infrastructure with a given dimension $\bar{w}$ during a given time period. In order to simplify the analysis, we will assume one homogenous user category of the infrastructure, that is, homogenous traffic of identical vehicles (with identical numbers of travellers) and identical time costs for all traffic participants.[7]

Let $k_t = k_t(Q_t, \bar{w}) = k_t(Q_t)$ denote the private (average) variable costs for the utilisation of an infrastructure with dimension $\bar{w}$ in period t. This includes the operating costs of the vehicle, the time costs of the trip, as well as the utilisation-dependent maintenance costs of the infrastructure. With increasing traffic, variable costs also increase.

Maximising social welfare of utilisation over the entire operating life T of a given transportation infrastructure results in the optimal congestion fee.[8]

$$\max_{Q_t} S = \sum_{t=1}^{T} \left[\int_0^{Q_t} P_t(\tilde{Q}_t) d\tilde{Q}_t - Q_t k_t(Q_t) \right] \qquad (3.10)$$

The first order condition is:

$$\frac{\partial S}{\partial Q_t} = P_t - \left[k_t(Q_t) + \frac{\partial k_t}{\partial Q_t} \cdot Q_t \right] = 0 \quad t = 1, \ldots, T \qquad (3.11)$$

From this follows:

$$P_t = k_t(Q_t) + \frac{\partial k_t}{\partial Q_t} \cdot Q_t \quad t = 1, \ldots, T \qquad (3.12)$$

The optimal congestion fee τ_t^0 can be calculated as the difference between gross price P_t and private variable costs $k_t(Q_t)$, that is:

---

[6] An overview of the literature on this topic can be found in Winston (1985, pp. 78 f.). The formal analysis in Mohring and Harwitz (1962) is still seminal, and the following simple model approach is based on it.

[7] This analysis can be generalised for the case of different categories of users utilising the same transportation infrastructure (e.g. Morrison, 1987).

[8] For simplification purposes the discount rate is disregarded.

$$\tau_t^0 = P_t - k_t(Q_t) = \frac{\partial k_t}{\partial Q_t} \cdot Q_t \quad t = 1, \ldots, T \tag{3.13}$$

which is equal to the (marginal) congestion costs of one utilisation of the given transportation infrastructure.

3.1.3.2 Congestion Fees with Variable Infrastructure

The level of the congestion costs depends on the dimension of the infrastructure. For instance, a certain level of traffic flow on a motorway with only two lanes will result in considerable traffic congestion, whereas on a four- or six-lane motorway this congestion will be considerably reduced or even disappear completely. High congestion fees on roads with high traffic volume due to increased traffic demand are an indication that new investment is necessary, with the optimal investment program depending on the level of congestion fees.

In the following it will be assumed that the infrastructure can be built without problems of indivisibility, so that differential calculus can be applied. The technical possibility to build infrastructures of different dimensions (capacities) and extend them gradually occurs frequently in practice (cf. Starkie, 1982). The bigger and more established the infrastructure, the lower the relevance of indivisibilities (cf. Keeler & Small, 1977). Important examples are motorways with several lanes or airports with several runways. In contrast, the following analysis cannot be applied to the marginal case of a pure public good where large technical indivisibilities cause minimal capacity to be so large in relation to demand that there is perfect non-rivalry in the utilisation of the infrastructure. For example, the difference between no road at all and the most simple form of a road constitutes a large technical indivisibility. Here, with a low grade of utilisation, externality costs disappear and the optimal congestion fee is 0. When building a transportation infrastructure, there can be either increasing, constant, or decreasing economies of scale in the relevant range of demand.

The price and investment problem when there are congestion costs can be formulated as follows (cf. Mohring & Harwitz, 1962; Keeler & Small, 1977): Let $\rho(w)$ denote the investment costs of the infrastructure with dimension w and an operating life of T periods (including the maintenance costs which are variable with road size). When traffic flow increases and the size of the infrastructure remains constant, the time costs of one trip increase: $\frac{\partial k_t}{\partial Q_t} > 0$. When the size of the infrastructure increases, the variable costs of a trip decrease: $\frac{\partial k_t}{\partial w} < 0$. Furthermore, it is assumed that $k_t(Q_t, w)$ is homogenous of degree 0 in Q_t and w. This means that traffic speed on the infrastructure depends only on the relation between traffic flow and dimension, and not on the absolute size of the infrastructure.[9]

---

[9] This assumption is typically made in the literature of transport economics when larger transportation infrastructures are concerned (e.g. motorways or airports) (cf. e.g. Mohring & Harwitz, 1962; Keeler & Small, 1977, p. 2).

3.1 Local (Path-Based) Externalities

The optimal price and investment rule is given by:

$$\max_{Q_t, w} S = \sum_{t=1}^{T} \left[\int_0^{Q_t} P_t(\tilde{Q}_t) d\tilde{Q}_t - Q_t k_t(Q_t, w) \right] - \rho(w) \quad (3.14)$$

The necessary conditions for the maximum result from differentiating of Eq. (3.14) with respect to Q_t and w.

The socially optimal pricing rule for one trip results from differentiating with respect to Q_t:

$$P_t = k_t(Q_t, w) + Q_t \frac{\partial k_t(Q_t, w)}{\partial Q_t}; \quad t = 1, \ldots, T \quad (3.15)$$

The first term after the equals sign denotes the private variable costs of one trip, the second term denotes the externality costs one additional vehicle imposes on all other vehicles. The price of one trip should thus be equal to the short-run social marginal costs of one trip.

The socially optimal investment rule results from differentiating with respect to w:

$$\rho'(w) = -\sum_{t=1}^{T} Q_t \frac{\partial k_t(Q_t, w)}{\partial w} \quad (3.16)$$

Simultaneously solving Eqs. (3.15) and (3.16) determines the optimal dimension of infrastructure w^o, as well as the optimal traffic flow Q_t^o. In particular, the following conclusions can be drawn:

- The price of one single trip must be equal to the short-run social marginal costs, including congestion costs. This necessitates the levying of an optimal congestion fee.
- The marginal costs of an additional unit of investment must be equal to the sum of the marginal benefits of all users of this infrastructure due to reduced variable costs.

3.1.4 Efficient Congestion Fees and Financing Objectives

So far, in the derivation of optimal price and investment rules, the financing problem has been disregarded. Nevertheless it is an interesting question to what extent optimal congestion fees can contribute to the recovery of investment costs (including the maintenance costs variable with infrastructure size).

3.1.4.1 The Contribution of Efficient Congestion Fees to the Financing of Transportation Infrastructures

The net revenues for financing a road (or another transportation infrastructure) can be calculated by multiplication of the optimal congestion fees with Q_t and summation over the total operating life[10]:

$$\sum_{t=1}^{T}[P_t(Q_t) - k_t(Q_t, w)]Q_t = \sum_{t=1}^{T} Q_t \frac{\partial k_t}{\partial Q_t} \cdot Q_t \quad (3.17)$$

Application of Euler's Theorem for homogenous functions of degree r to the function $k_t(Q_t, w)$ results in:

$$r \cdot \sum_{t=1}^{T} k_t(Q_t, w) = \sum_{t=1}^{T} Q_t \frac{\partial k_t}{\partial Q_t} + \sum w \frac{\partial k_t}{\partial w} \quad (3.18)$$

Applied for functions k_t homogenous of degree 0:

$$\sum_{t=1}^{T} Q_t \cdot \frac{\partial k_t}{\partial Q_t} = \sum_{t=1}^{T} -w \frac{\partial k_t}{\partial w}; \quad t = 1, \ldots, T \quad (3.19)$$

Thus, due to the optimal investment rule (3.16) it follows that:

$$\sum_{t=1}^{T} Q_t \frac{\partial k_t}{\partial Q_t} \cdot Q_t = w\left[-\sum_{t=1}^{T} \frac{\partial k_t}{\partial w} Q_t\right] = w\rho'(w) \quad (3.20)$$

And:

$$\sum_{t=1}^{T}[P_t(Q_t) - k_t(Q_t, w)]Q_t = w\rho'(w) \quad (3.21)$$

If there are constant economies of scale for the expansion of the infrastructure (or if existing economies of scale are exhausted), it holds that: $w\rho'(w) = \rho(w)$. In that case the investment costs for the optimal infrastructure capacity are exactly covered by the optimal congestion fees. If there are increasing economies of scale for the expansion of the infrastructure, it holds that: $w\rho'(w) < \rho(w)$, and levying optimal congestion fees leads to a deficit. If the expansion of the infrastructure is associated with decreasing economies of scale (for instance because of long-run increasing marginal costs for road building), optimal congestion fees lead to a surplus (cf. Mohring & Harwitz, 1962, pp. 81–86).

---

[10] For simplification purposes the discount rate is set at zero.

3.1.4.2 Socially Optimal Congestion Fees Under Cost-Covering Constraint

If there are increasing economies of scale for the expansion of a transportation infrastructure, and thus socially optimal congestion fees cannot recover all investment costs, the problem arises how to cover the remaining deficit. An alternative to covering the remaining deficit with money from the public budget is to levy welfare-maximising linear fees under the cost-covering constraint, which will be examined in the following. For the case of a (short-run) given infrastructure level, this problem corresponds to the derivation of the well-known Ramsey prices (cf. Baumol & Bradford, 1970; Ramsey, 1927).[11]

When there is perfect competition with many active suppliers, firms will set prices equal to marginal costs. If there are increasing economies of scale, marginal cost pricing will not cover the total costs of production. From a welfare economic point of view it is then important to raise prices exactly so far above marginal costs that total cost is covered, but that the resultant loss of consumer surplus in total is as small as possible. The concept of Ramsey prices is based on the assumption that it is possible to split total demand for infrastructure capacities into several groups in such a way that for each group one particular uniform price can be enforced.

In the following we consider an infrastructure with homogeneous capacity without quality of service differentiation. The Ramsey pricing rule requires that prices on markets with price elastic (sensitive) demand are raised relatively slightly, while on markets with inelastic demand relatively high prices can be charged. This can be easily comprehended intuitively, because in general the loss of consumer surplus due to a price increase is the greater, the more elastic the reaction of demand to the price increase.[12]

Maximising consumer and producer surplus under the cost-covering constraint results in second-best access tariffs. In doing so, mark-ups corresponding to the elasticities in demand for transport capacities are to be charged. The lower the price elasticity, the higher the necessary mark-up on the marginal costs of infrastructure use (cf. e.g. Baumol & Bradford, 1970).

Linear access tariffs according to the Ramsey price principle constitute a theoretical reference point for socially optimal network access fees. In competitive markets the market process can generate such a price structure by taking into account different demand elasticities resulting in endogenous Ramsey pricing (cf. Willig & Baumol, 1987, p. 34). If there are externality costs, mark-ups (corresponding to demand elasticities) on social marginal costs are to be charged (cf. Morrison, 1987, pp. 47f.). In the following, however, second-best congestion fees in case of variable infrastructure levels are to be derived. The question of

---

[11] Possible alternatives are peak load pricing, taking into account temporary fluctuations in capacity utilisation, as well as multi-part tariffs (a basic fee plus an additional utilisation-dependent price component) (cf. Chap. 4).

[12] Ramsey prices are a form of price differentiation. For the concept of price differentiation see more explicitly Chap. 4.

interest in this case is if, and to what extent, the investment rule changes compared to the first-best solution.[13]

We will examine user categories (or user classes), which are different with regard to the private costs of one trip, the externality costs, as well as the price elasticities of demand for transportation infrastructure capacities.[14]

Let $P_{it}(Q_{it})$ denote the inverse demand function of user category i $(i = 1,...,n)$ in period t $(t = 1,...,T)$; P_{it} is the "gross price" of one trip of one traffic participant of category i in period t and Q_{it} denotes the number of trips of traffic participants i during period t. $k_{it} = k_{it}(Q_{1t},...,Q_{nt})$ denotes the private costs of one traffic participant of category i in period t. $\rho(w)$ denotes the investment costs of the transportation infrastructure with dimension w and with an operating life of T periods, which have to be financed from the congestion fees. For simplification purposes it is assumed that demand is independent between the different user categories and between the time periods and that the maintenance costs of the infrastructure are negligible. In the following, optimal congestion fees and the optimal dimension of the infrastructure will be derived simultaneously under the cost-covering constraint.

Maximisation of social welfare under the cost-covering constraint is:

$$\max_{(Q_{it},...,Q_{nt},w)} S = \sum_{t=1}^{T}\left[\sum_{i=1}^{n}\int_{0}^{Q_{it}} P_{it}(\widetilde{Q}_{it})d\widetilde{Q}_{it} - \sum_{i=1}^{n} k_{it}(Q_{1t},...Q_{nt},w)Q_{it}\right]$$
$$- \rho(w) \qquad (3.22)$$

in such a way that

$$\sum_{t=1}^{T}\sum_{i=1}^{n}[P_{it} - k_{it}(Q_{1t},...,Q_{nt},w)]Q_{it} = \rho(w) \qquad (3.23)$$

---

[13] This question also arises, if there are decreasing economies of scale for the expansion of a transportation infrastructure (e.g. for airports) and tariffs based on marginal costs would lead to a surplus. The following analysis can be applied to this case analogously.

[14] Optimal user fees can be generalised analogously for the case of different categories of users utilising the same transportation infrastructure. This is advisable, if different categories of users (for instance trucks and cars) cause different externality costs.

3.1 Local (Path-Based) Externalities

The Lagrange function is:

$$L(Q_{1t},\ldots,Q_{nt},w,\lambda) = \sum_{t=1}^{T}\left[\sum_{i=1}^{n}\int_{0}^{Q_{it}} P_{it}(\tilde{Q}_{it})d\tilde{Q}_{it} - \sum_{i=1}^{n}k_{it}(Q_{1t},\ldots Q_{nt},w)Q_{it}\right] - \rho(w)$$
$$+\lambda\left[\sum_{t=1}^{T}\sum_{i=1}^{n}[P_{it}-k_{it}(Q_{1t},\ldots,Q_{nt},w)]Q_{it} - \rho(w)\right]$$

(3.24)

$$MC_{it}^{S}(Q_{1t},\ldots,Q_{nt},w) = \frac{\partial}{\partial Q_{it}} \cdot \sum_{j=1}^{n} k_{jt}(Q_{1t},\ldots,Q_{nt},w) \cdot Q_{jt}$$
$$= k_{it}(\cdot,w) + \sum_{j=1}^{n} \frac{\partial k_{jt}(\cdot,w)}{\partial Q_{it}} \cdot Q_{jt} \quad (3.25)$$

denotes the social marginal costs of one additional vehicle of category i in period t with an infrastructure capacity w. This includes the externality costs imposed on all other vehicles (vehicles in all other categories j, as well as vehicles in the same category i).

$$MR_{it} = \frac{\partial}{\partial Q_{it}}[P_{it}(Q_{it}) \cdot Q_{it}] \quad (3.26)$$

denotes the marginal revenue of the price of one trip in user category i during period t.

Then it is true that:

$$\frac{\partial L}{\partial Q_{it}} = (P_{it} - MC_{it}(\cdot,w)) + \lambda[MR_{it} - MC_{it}(\cdot,w)] = 0 \quad (3.27)$$

$$\frac{\partial L}{\partial w} = \sum_{t=1}^{T}\left[-\sum_{i=1}^{n}\frac{\partial k_{it}}{\partial w}\cdot Q_{it}\right] - \frac{\partial \rho(w)}{\partial w} + \lambda\left[\sum_{t=1}^{T}\sum_{i=1}^{n} -\frac{\partial k_{it}}{\partial w}Q_{it} - \frac{\partial \rho}{\partial w}\right] = 0 \quad (3.28)$$

From Eq. (3.27) it follows that:

$$\frac{P_{it} - MC_{it}(\cdot,w)}{MC_{it}(\cdot,w) - MR_{it}} = \lambda \quad (3.29)$$

$$MR_{it} = P_{it}\left[1 + \frac{\partial P_{it}}{\partial Q_{it}} \cdot \frac{Q_{it}}{P_{it}}\right] = P_{it}[1 + \varphi_{it}] \quad (3.30)$$

Therefore it is true that:

$$\frac{P_{it} - MC_{it}(\cdot, w)}{MC_{it}(\cdot, w) - [P_{it}(1 + \varphi_{it})]} = \lambda \qquad (3.31)$$

and consequently:

$$\frac{P_{it} - MC_{it}(\cdot, w)}{P_{it}} = -\frac{\lambda}{1 + \lambda} \cdot \varphi_{it} \qquad t = 1, \ldots, T; \qquad i = 1, \ldots, n \qquad (3.32)$$

From Eq. (3.28) it follows that:

$$\sum_{t=1}^{T}\left[-\sum_{i=1}^{n}\frac{\partial k_{it}}{\partial w} \cdot Q_{it}\right] - \frac{\partial \rho}{\partial w} + \lambda\left[\sum_{t=1}^{T}\sum_{i=1}^{n} -\frac{\partial k_{it}}{\partial w} \cdot Q_{it} - \frac{\partial \rho}{\partial w}\right] = 0 \qquad (3.33)$$

And thus:

$$(1 + \lambda)\left(-\frac{\partial \rho}{\partial w}\right) = (1 + \lambda)\sum_{t=1}^{T}\sum_{i=1}^{n}\left(\frac{\partial k_{it}}{\partial w} \cdot Q_{it}\right) \qquad (3.34)$$

The optimal price and investment rules under the cost-covering constraint result from Eqs. (3.32) and (3.34) as follows:

$$\rho'(w) = -\sum_{t=1}^{T}\sum_{i=1}^{n}\frac{\partial k_{it}(Q_{1t}, \ldots, Q_{nt}, w)}{\partial w} \cdot Q_{it} \qquad (3.35)$$

$$\frac{P_{it} - \left[k_{it}(\cdot, w) + \sum_{j=1}^{n}\frac{\partial k_{jt}(\cdot, w)}{\partial Q_{it}} \cdot Q_{jt}\right]}{P_{it}} = \frac{-\lambda}{1 + \lambda} \cdot \varphi_{it} \qquad (3.36)$$

$$t = 1, \ldots, T; \qquad i = 1, \ldots, n$$

Here φ_{it} denotes the quantity elasticity of the inverse demand function of user category i during period t.

From Eqs. (3.35) and (3.36) it follows:

- Analogous to the case without cost-covering constraint, the investment level must be expanded until the marginal costs of one additional unit of investment are equal to the marginal utility (particularly due to shorter travel time).
- The second-best congestion fee is given by a mark-up on the social marginal costs. The higher the quantity elasticity of the inverse demand function (that is, the lower the price elasticity of the relevant user category), the higher the necessary mark-up.

3.1 Local (Path-Based) Externalities

– The simultaneous solution of Eqs. (3.35) and (3.36) determines the second-best infrastructure dimension w^{**} as well as the second-best traffic flow Q_{it}^{**}; $i=1,\ldots,n$, $t=1,\ldots,T$ (with second-best congestion fees). In the same way, the second-best congestion fee $\tau_{it}^{**} = P_{it} - k_{it}(\cdot, w^{**})$ is determined as the deviation from social marginal costs

$$k_{it}(\cdot, w^{**}) + \sum_{j=1}^{n} \frac{\partial k_{jk}(\cdot, w^{**})}{\partial Q_{it}} \cdot Q_{jt}$$

with second-best infrastructure level. Thus, although the investment rule is the same as in the case without the cost-covering constraint, a different capacity level results.

3.1.5 Congestion Externalities and Quality Differentiation in Infrastructure Networks

So far, in order to demonstrate the basic principle of congestion externalities and optimal congestion fees, we have examined the traffic flow on one individual road (or on one airport, or one railway track) in isolation. However, transportation infrastructures are typically located in a network of other infrastructures, which can be used by traffic participants in a complementary or substitutive way. The traffic flow on a motorway can thus also depend on the supply of substitutive bypass roads. Airplanes starting on a particular airport need a complementary landing slot on the destination airport, etc.

However, irrespective of these relations of substitution and complementarity from the perspective of individual traffic participants, traffic flows on the different infrastructures are not systematically and causally connected. The congestion externalities on a given infrastructure depend exclusively on the traffic flow on this specific infrastructure. They constitute local congestion externalities, because the opportunity costs for using, for example, one particular section of motorway can be limited to this section of motorway.

The question of how many vehicles in total are using the parallel roads must be distinguished from the question of how traffic flows are split up onto parallel infrastructures. Price and investment decisions for the case of substitutive and complementary highways have been the subject of substantial theoretical and empirical research in transport economics in the last decades.[15] Many studies relate to a scenario with two aligned (parallel) roads and one complementary joint road (cf. Verhoef & Small, 2004, p. 131).

Congestion problems in infrastructure networks can be analyzed, taking into account basic substitutive and complementary relations, with the help of simple stylised traffic structures. In doing this, graph theory can be useful. The nodes of the

---

[15] For an overview, cf., e.g., Santos (ed., 2004), Verhoef and Small (2004), Mohring (1999).

Fig. 3.5 Parallel paths

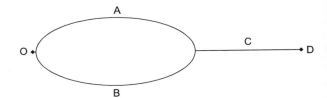

graph represent crossings, while the links (edges) represent (parallel) separate paths, or instead separate roads, so that the traffic flows occur and are measured separately (isolated) on the individual links.

3.1.5.1 Spontaneous Symmetrical Traffic Distribution in the Absence of Road Quality Differentiation

The underlying scenario consists of a point of origin O and a point of destination D, which are connected by two parallel paths A and B with separate, aligned traffic flows (separate roads or parallel, separated lanes of a motorway), as well as one joint path C. This scenario is illustrated by Fig. 3.5 (cf. Verhoef & Small, 2004, p. 131, Fig. 1).

In the language of graph theory these are two routes: Route AC, consisting of links A and C, and Route BC, consisting of links B and C. Points O and D can be interpreted as nodes of the graph, which function as points of access and departure, as well as road crossings. It is assumed that quality (including technical characteristics, e.g. paving etc.) as well as capacity (maximum traffic flow) of paths A and B are identical. On both paths A and B the identical traffic flow $\frac{Q^o}{2}$ occurs spontaneously, so that the total travel time will be minimised on each path. Incentives for traffic participants to use path A instead of B (or vice versa) do not develop, because after changing from A to B the variable costs of one trip will increase, because $k\left(\frac{Q^o}{2}+1\right) \geq k\left(\frac{Q^o}{2}\right)$. Congestion fees are not necessary either in this case for realising the time-cost minimising symmetrical distribution of vehicles $Q_A = Q_B = \frac{Q}{2}$ on the two parallel roads. Diverging from this "optimal" solution would not be incentive compatible from an individual perspective either, because individual travel costs would increase.

3.1.5.2 The Braess Paradox: The Role of Different Road Qualities

The possibility of the Braess paradox is based on the assumption that although congestion externalities are identical for the sum of sections $A = a_1 + a_2$ and $B = b_1 + b_2$, they are opposite on the individual sections, so that $a_1 = b_2$ (low) and $a_2 = b_1$ (high). This means that sections a_1, b_2 are of higher quality than sections a_2 and b_1. A travel-cost minimising spontaneous distribution of traffic flows is no longer guaranteed if paths A and B are linked by an intersecting road as shown in Fig. 3.6. This phenomenon has become well-known in the literature on traffic planning through the so-called Braess paradox (cf. Braess, 1968). This paradox states that a situation cannot be excluded where, when a traffic network is expanded

3.1 Local (Path-Based) Externalities

Fig. 3.6 Braess paradox

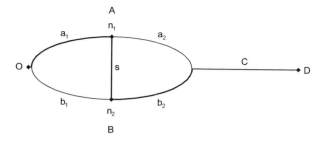

by one additional road, the optimal (overall time-cost minimising) traffic flows are no longer incentive compatible, because some individual drivers can work out more time-saving routes. When the additional road is then blocked, traffic flows again divide in a more advantageous way, so that in this case the total time expenditure of the traffic flows is reduced by a reduction of the road network. The model example with four nodes (cf. Braess, 1968, p. 263) can be illustrated by expanding Fig. 3.5 (cf. Fig. 3.6).

Before the intersecting road s was built, the only choice was between the two routes (a_1, a_2, C) and (b_1, b_2, C).[16] After the intersecting road s was built, the two additional routes (a_1, s, b_2, C) and (b_1, s, a_2, C) are available. Braess examines the question, whether by building the intersecting road s the time-cost minimising symmetrical distribution of the traffic flows on the two routes (AC) and (BC) can be subverted by some individual drivers using the intersecting road and choosing routes (a_1, s, b_2, C) or (b_1, s, a_2, C). If this is possible, a paradoxical situation results, where building an additional intersecting road causes a slowing down of traffic flows, so that total travel time increases. Under the assumption of identical quality of the parallel roads A and B, before the intersecting road is built, there are symmetrical congestion conditions on A and B, and thus also on sections a_1, a_2 and b_1, b_2 (as traffic flows on A and B remain unchanged). Because at the optimum, before the intersecting road is available, traffic flows on sections b_2, a_2 are identical, there is no incentive for an individual driver to use the intersecting road. The additional time costs of using intersecting road s are not balanced by saving time on section b_2. On the contrary, by changing to b_2, congestion externalities are marginally reduced on a_2, and marginally increased on b_2. The Braess paradox does not occur.

Under the preconditions of quality differentiation, so that a_1, b_2 (low) and a_2, b_1 (high) there is an incentive for one individual trip through (a_1, s, b_2, C), to avoid the optimal, time-cost minimising route (a_1, a_2, C), because the lower congestion costs on b_2 (in spite of the marginal increase of congestion on b_2) as compared to the high congestion costs on a_2 make the detour via intersecting road s advantageous.

---

[16] A route is defined as a combination of sections of road that connect starting point O to destination D.

For transportation practice the question of the relevance of the Braess paradox arises. Different congestion conditions on one path are possible, if capacity, quality of the paving etc. differ on the individual sections, so that a_1, b_2 have a higher quality and a_2, b_1 a lower quality. Intersecting road s offers the possibility of bypassing routes (a_1, a_2, C) and (b_1, b_2, C). If all vehicles choose route (a_1, s, b_2, C), total travel time will be increased compared to the former situation without the intersecting road (cf. Cohen & Horowitz, 1991, p. 701). However, it would be a mistake to conclude from this that infrastructure expansion in traffic networks is counterproductive. Instead it is necessary to introduce congestion fees in order to control traffic flows. Samuelson (1992, p. 7) already proves that the Braess paradox disappears, when suitable congestion fees are charged, because traffic flows will distribute at a ratio of ¼ to $^3/_8$ to $^3/_8$ on the routes (a_1, s, b_2, C), (a_1, a_2, C) and (b_1, b_2, C). The most expensive, but also the fastest link is route (a_1, s, b_2, C).

Braess' analysis is based on the planning approach of minimising time costs of traffic flows over all routes. In the following it will be shown that suitable congestion fees and the resulting quality differentiation between parallel paths lead to welfare improvements, even though they do not necessarily minimise total travel time.

3.1.5.3 Congestion Fees and Infrastructure Quality Differentiation

The Pigou–Knight controversy already focused on the case of two parallel roads, with drivers being able to choose between road A with low infrastructure quality but unlimited capacity, and road B with high infrastructure quality but limited capacity. It was beyond dispute in the Pigou-Knight controversy that this choice had to be determined according to the social marginal costs and not the (average) variable costs. This is illustrated by the following Fig. 3.7. Because of the high capacity, on road A the private variable costs of usage are constant and equal to the social marginal costs. On road B, due to the higher quality of the infrastructure, the variable costs are lower as long as the road is not congested, but with increasing congestion, both the private and the social costs increase, because each additional vehicle lifts the costs of all other vehicles to the level of the additional vehicle (cf. Knight, 1924, p. 588; Pigou, 1920, Appendix iii).

Q^o denotes the socially optimal traffic flow, Q^e denotes the traffic flow without congestion fees on road B. Due to perfect non-rivalry no congestion fee is necessary on road A. The actual controversy centred on the question whether the necessary congestion fee should be levied through a tax or through a profit-maximising congestion fee. It is Pigou who deserves the credit for working out the necessity of adapting drivers' individual decision-making to social marginal cost. For the case of a public infrastructure, this problem can be solved through a usage-dependent special tax (Pigouvian tax) that is equal to congestion costs. For the case assumed by Knight, of a profit-maximising road owner in effective competition (to an alternative, competing road), the profit-maximising price is equal to the social marginal costs (Eq. 3.6); from this the socially optimal congestion fee τ^o results (cf. Knight, 1924, p. 587). This case already illustrates the quality differentiation caused by congestion fees. In the social optimum, vehicles on road B will, in

3.1 Local (Path-Based) Externalities

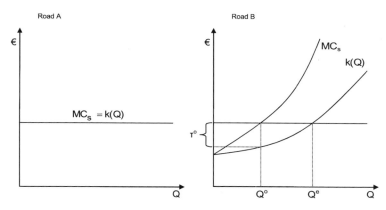

Fig. 3.7 Competition between parallel roads

exchange for paying a congestion fee, arrive at their destination faster (at lower k(Q)) than vehicles on road A without a congestion fee but with higher time expenditure. Consequently, the efficient gross price of one trip is identical on both roads.

3.1.5.4 Congestion Fees and Quality of Service Differentiation

With parallel roads of identical quality, traffic will distribute symmetrically, as long as the congestion fees for both roads are identical.[17] However, parallel roads offer the possibility of differentiating congestion fees, with the aim of differentiating quality by means of different traffic flows. If different congestion fees ($\tau_A > \tau_B$) are charged on the two identical roads A and B, traffic will no longer divide symmetrically. Traffic participants with high time preference will choose road A and pay the high congestion fee τ_A, while traffic participants with low time preference will choose road B with the lower congestion fee τ_B.[18]

We will examine the extreme case, where only two user groups Q_A and Q_B can be differentiated, with vehicles in traffic flow Q_A having high time preference for travelling without congestion externalities, and vehicles in traffic flow Q_B having no high time preference. Optimal congestion fees can be derived with the help of Fig. 3.8. While there is a very high congestion fee on road A, on road B the congestion fee is low.

Recently, various empirical studies have been conducted on the distribution of traffic participants' preferences regarding travel time and reliability, which prove that there is considerable heterogeneity in the way that travel time and reliability are evaluated. Congestion fee models which take this heterogeneity into account can achieve considerable welfare improvements (e.g. Small, Winston, & Yan, 2005).

---

[17] The same holds true for different parallel lanes, as long as they are separated.

[18] The formal derivation of the socially optimal congestion fees for express road A and standard road B can be found in Verhoef and Small (2004, pp. 133–137).

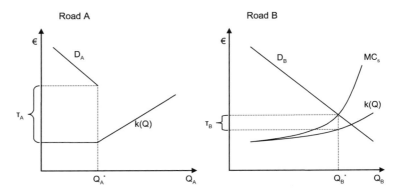

Fig. 3.8 Uncongested travel on road A

3.1.6 Congestion Fees in a Monopoly

The Pigou–Knight controversy already illustrated that a profit-maximising infrastructure provider internalises congestion externalities in his congestion fee. However, this was demonstrated for the case of competition between two parallel roads (cf. Fig. 3.7). When infrastructures are provided in a monopoly scenario (with both active and potential competition being absent), the question arises whether infrastructure usage fees in a monopoly also take congestion externalities into account.[19] First it is important to remember that even a monopolist who possesses the property rights to charge congestion fees has to take congestion externalities into account, because they reduce the fee which he can demand from the infrastructure users (cf. Edelson, 1971, pp. 874f.). Therefore congestion externalities as (negative) consumer externalities are fundamentally different from production externalities (noise, pollution, etc.), which cannot be attributed directly to beneficiaries (consumers) of the production and thus can be disregarded by the producer.[20]

For the case of a homogenous traffic flow Q_t, $t = 1,\ldots,T$ and a (short-run) given infrastructure level $\overline{w}$, in a monopoly the profit-maximising traffic flow Q_t^m results as follows:

$$\max_{Q_t} \pi = \sum_{t=1}^{T} \left(P_t(Q_t) \cdot Q_t \right) - Q_t k_t(Q_t, \overline{w}), \tag{3.37}$$

where marginal revenue MR_t must be equal to social marginal costs $MC_{s,t}$.

---

[19] These are monopolistic bottleneck facilities. If, irrespective of market type, congestion externalities are taken into account, market power regulation can be limited to price level regulation, cf. Chap. 8.

[20] Buchanan shows that levying a Pigou tax in order to internalise external effects of production can decrease welfare in case of a monopoly (second-best fallacy), cf. Buchanan (1969).

3.1 Local (Path-Based) Externalities

$$MR_t = P_t + \frac{\partial P_t}{\partial Q_t} \cdot Q_t = k_t + \frac{\partial k_t}{\partial Q_t} \cdot Q_t = MC_{s,t} \quad t = 1, \ldots, T \quad (3.38)$$

with an accompanying gross monopoly price P_t^m.
It holds that:

$$\frac{P_t^m - MC_{s,t}}{P_t^m} = -\frac{1}{\varepsilon_t} \quad \text{with} \quad \varepsilon_t = \frac{\partial Q_t}{\partial P_t} \cdot \frac{P_t}{Q_t} \quad (3.39)$$

Consequently the operator of a private road infrastructure in a monopoly receives a profit-maximising utilisation fee:

$$\tau_t^m = P_t^m - k_t = \frac{\partial k_t}{\partial Q_t^m} \cdot Q_t^m + a_t^m \quad (3.40)$$

where $a_t^m = -\frac{\partial P_t}{\partial Q_t^m} \cdot Q_t^m$ corresponds to the monopoly mark-up. The lower the price elasticity of the demand function, the higher the necessary mark-up on social marginal costs. This shows that only orientation to social marginal costs leads to the profit-maximising price structure. Thus externality costs are used for determining usage fees not only under competition, but also in a monopoly situation. Figure 3.9 illustrates how the monopolistic congestion fee is derived; for simplification purposes, time index t is omitted.

The price and investment decision of a monopolistic infrastructure provider results as follows[21]:

$$\max_{Q_t, w} \pi = \sum_{t=1}^{T} [P_t(Q_t)Q_t - Q_t k_t(Q_t, w)] - \rho(w) \quad (3.41)$$

Differentiating with respect to Q_t results in the monopoly pricing rule P_t^m and the associated monopoly utilisation fee:

$$\tau_t^m = P_t^m - k_t. \quad (3.42)$$

Differentiating with respect to w results in the optimal investment rule:

$$w^m \rho'(w^m) = -\sum_{t=1}^{T} w^m \frac{\partial k_t(Q_t^m, w^m)}{\partial w^m} \cdot Q_t^m \quad (3.43)$$

Analogous to the case of welfare maximisation (with or without the cost-covering constraint), the investment level must be expanded to such an extent that the

---

[21] Under identical basic assumptions as in the welfare maximising case (Sect. 3.1.3).

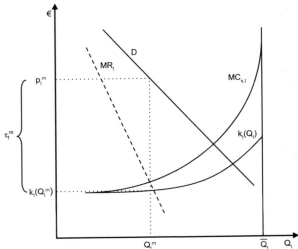

Fig. 3.9 Congestion fees in a monopoly

marginal costs of one additional investment unit are equal to the marginal utility (through shorter travelling time).

The simultaneous solution of the monopoly pricing rule (3.42) and the investment rule (3.43) determines the profit-maximising infrastructure dimension w^m as well as the profit-maximising traffic flow Q_t^m, $t=1,\ldots,T$ (for monopolistic user fees). The monopolistic user fee $\tau^m = P_t^m - k_t(Q_t^m, w^m)$ for monopolistic infrastructure level w^m is determined analogously.

Although the investment rule does not differ from the welfare-maximising investment rule, the resulting capacity level is different. A comparison between the infrastructure level in a monopoly and the socially optimal infrastructure level shows that the profit-maximising infrastructure level in a monopoly is smaller than the socially optimal infrastructure level. This can be explained as follows:

Due to the optimal investment rule, for optimal infrastructure level w^o and socially optimal traffic flow Q^o it holds that:

$$\sum_{t=1}^{T} Q_t^o \frac{\partial k_t}{\partial Q_t^o} \cdot Q_t^o = w^o \rho'(w^o) \qquad (3.44)$$

(cf. Eq. 3.20) and thus the socially optimal dimension:

$$w^o = \frac{\sum_{t=1}^{T} Q_t^o \cdot \frac{\partial k_t}{\partial Q_t^o} \cdot Q_t^o}{\rho'(w^o)} \qquad (3.45)$$

In a monopoly it is true for all periods that traffic flow Q^m is smaller than socially optimal traffic flow Q^o, and thus externality costs are also lower if the socially

optimal infrastructure dimension w^o (instead of the optimal monopolistic dimension) is utilised.
Therefore

$$w^o = \frac{\sum_{t=1}^{T} Q_t^o \cdot \frac{\partial k_t(\cdot, w^o)}{\partial Q_t^o} \cdot Q_t^o}{\rho'(w^o)} > \frac{\sum_{t=1}^{T} Q_t^m \cdot \frac{\partial k_t(\cdot, w^o)}{\partial Q_t^m} \cdot Q_t^m}{\rho'(w^o)} \qquad (3.46)$$

The socially optimal infrastructure is thus overdimensioned for the lower monopolistic traffic flow Q^m. For the optimal investment rule to be fulfilled in the monopoly case, it must be true that $w^m < w^o$ where

$$w^m = \frac{\sum_{t=1}^{T} Q_t^m \cdot \frac{\partial k_t(Q_t^m, w^m)}{\partial Q_t^m} \cdot Q_t^m}{\rho'(w^m)}. \qquad (3.47)$$

This can be shown as follows: Under application of Euler's theorem for k_t homogenous of degree 0 the profit maximising investment rule (3.43) results in:

$$w^m \rho'(w^m) = \sum_{t=1}^{T} Q_t^m \frac{\partial k_t(Q_t^m, w^m)}{\partial Q_t^m} \cdot Q_t^m \qquad (3.48)$$

3.1.7 Congestion Fees in Traffic Practice

In the last decades the charging systems in the EU Member States and in Switzerland has been characterised by increasing user financing. In the period from the 1950s up until the present there has been a general shift from budgetary financing to user financing. In this context an increasing transition from a time-dependent area toll (vignette) to a mileage-based toll can be observed (cf. Baumgarten, Huld, & Hartwig, 2013, pp. 92ff.).

3.1.7.1 Road Usage Fees and HGV (=Heavy Goods Vehicle) Toll

With the introduction of the HGV toll the issue of usage fees for transportation infrastructures has gained relevance in Germany.[22] From the perspective of the Trans-European road network, the EU has supplied important impulses for levying road tolls. In July 1998 the European Commission published a White Paper on this topic (COM (98) 466, July 1998, "Fair payment for infrastructure use: A phased approach to a common transportation infrastructure charging framework in the EU"), which has become the object of numerous controversies. The 1999 "Directive on the charging of heavy goods vehicles for the use of certain infrastructures"

---

[22] Cf. Wissenschaftlicher Beirat beim Bundesministerium für Verkehr, Bau, und Wohnungswesen (1999, 2000, 2005); Knieps (2006); Wieland (2005).

already contained essential elements for the design of road toll and usage fees; it was, however, limited to motorways and roads similar to motorways.[23] It states that average road tolls have to be geared toward recovering the costs for building, maintaining and expanding the road network in question. Differentiating for time of day and environmental standards of the vehicles is permissible. In August 2003, the European Commission introduced a proposal for amending this Directive, with the aim of making it easier to include subordinated parts of the road network (expressways) and light trucks (over 3.5 t).[24] A further concern is better differentiation in levying road tolls, for instance by type of vehicle, time or location. The reform process towards usage fees taking into account pollution and congestion caused by actual use of vehicles is ongoing.[25]

The 1998 EU White Paper recommended charging infrastructure usage fees according to social marginal costs, leaving the problem of total cost recovery of transportation infrastructures to chance (European Commission, 1998). The 1999 Directive focused on the problem of cost recovery. In the 2003 proposal the question of differentiated usage fees for recovering costs moved to the foreground. Thus in European traffic infrastructure policy a gradual movement from tax-based to user-based financing can be observed.

Road cost calculation (or more general infrastructure accounting) is a well-known topic in transportation policy; recently, however, they received new impulses in the context of designing contracting schemes for the private financing of transportation infrastructure investments. The necessary private capital can only be acquired when the accompanying entrepreneurial risk (utilisation and revenue risk, etc.) is compensated through a market-compliant interest rate. This leads inevitably to the necessity of forward-looking cost accounting, which has to be based on a decision-based determination of capital costs (cf. Chap. 2). Infrastructure costs have to be differentiated from environmental protection and traffic safety costs.

The optimal usage fee reflects the short-run marginal costs, including congestion costs, and thus is identical to the opportunity costs of infrastructure usage. As different user groups (trucks, cars) cause different congestion costs, the necessity for different congestion fees is a logical consequence. It is necessary to differentiate between the different user categories (or user classes) with different private costs for one trip (including different time costs), different congestion costs, as well as different price elasticities of demand for road infrastructure capacities. As the

---

[23] Directive 1999/62/EC of the European Parliament and of the Council of 17 June 1999 on the charging of heavy goods vehicles for the use of certain infrastructures OJ L 187, 20.07.1999, pp. 42–50.

[24] Proposal for a Directive of the European Parliament and of the Council amending Directive 1999/62/EC on the charging of heavy goods vehicles for the use of certain infrastructures, COM/2003/0448 final, 29.08.2003.

[25] Directive 2011/76/EU of the European Parliament and of the Council of 27 September 2011 amending Directive 1999/62/EC on the charging of heavy goods vehicles for the use of certain infrastructures (OJ L 269/1, 14.10.2011).

opportunity costs for utilising scarce transportation infrastructure capacities depend on all user groups, all user groups have to be included when charging user fees, that is, not only heavy trucks, but also light trucks, buses and cars on the roads. Insofar as different user groups cause different congestion costs, it follows that different usage fees must be charged.

3.1.7.2 Inner City Toll

In European countries, road tolls in the form of vignettes valid for certain periods of time, or utilisation-dependent charges have in the past mostly been levied in order to finance road infrastructures. However, due to increasingly pressing congestion problems, the issue of traffic management by means of congestion fees has become more important in traffic policy reforms. While the introduction of an inner city toll in Bergen (1996), Oslo (1990) and Trondheim (1991) was primarily intended to help finance road infrastructure, the aim of the London city toll, implemented in 2003, was traffic management and a reduction of congestion. In the first 6 months after implementation traffic decreased by 15 % on average (cf. Schade, 2005). Since 2003 vehicles driving into the London city centre on work days between 7 a.m. and 6 p.m. have had to pay a road toll of £ 5,--, raised to £ 8,-- in 2005 and £ 10,– in 2011. There are various exceptions, for example licensed taxis or busses; residents get a 90 % reduction. This is a fee to be paid for driving inside the chargeable area during a work day, with no higher charge during rush hour and irrespective of the distance covered within the chargeable area. The introduction of the Stockholm inner city toll also has the objective of managing traffic and reducing congestion problems. However, in Stockholm peak load pricing has been implemented in the form of time-of-day dependent charges, with the highest charges to be paid at peak times (cf. Müller-Jentsch, 2013).

3.1.7.3 Chargeable Expressways

Since the 1991 Intermodal Surface Transportation Efficiency Act (ISTEA) and the 1998 Transportation Equity Act for the twenty-first century (TEA-21) the levying of utilisation-dependent road usage fees has become increasingly important in the USA. Chargeable High Occupancy Toll (HOT) roads constitute a variant of the High Occupancy Vehicle (HOV) roadways. While on High Occupancy Vehicle roadways only vehicles with at least 2 occupants are allowed, this user constraint does not exist on High Occupancy Toll roads and is replaced by a congestion fee (cf. Lindsey, 2005, p. 49). As the number of persons in a car has no impact on the congestion costs, the introduction of High Occupancy Toll congestion fees is an improvement over the High Occupancy Vehicle constraints. Thus there is considerable potential for achieving welfare improvements by means of applying suitable price differentiation schemes, both for express lanes and standard lanes. In doing so, taking the heterogeneity of users with regard to duration and reliability of travel time into consideration is of particular importance (e.g. Small et al., 2005, 2006; Verhoef & Small, 2004).

In the meantime, various pilot projects have been completed with the aim of establishing quality differentiation between aligned roadways or parallel lanes:

chargeable express roads with low traffic flow and free roadways with high traffic flow. Chargeable High Occupancy Toll roads were introduced in Los Angeles, San Diego, Houston and Minneapolis (for example the Interstate System Construction Toll Pilot Program and the Express Lanes Demonstration Program). In the meantime some results from pilot projects are available, regarding the feasibility of implementing, by means of suitable electronic toll collection systems, congestion fees that are dependent on the time of day and spatially differentiated. In its comprehensive final report (2009) the US National Surface Transportation Infrastructure Financing Commission comes to the conclusion that usage-dependent congestion fees are far superior to all conceivable alternatives for financing road infrastructure. As optimal congestion fees allow a more efficient utilisation of existing infrastructure, the requirement for additional capacities is reduced. The Commission advocates a consistent adoption of utilisation-dependent toll systems despite significant implementation problems (administrative costs, bypass behaviour, personal data protection concerns and public acceptance). The Commission also points to the huge technological progress in traffic telematics which has lead to a strong decrease in the costs of toll collection systems over the last decade (cf. U.S. National Surface Transportation Infrastructure Financing Commission, 2009, pp. 125ff.). A fundamental extension of the legal basis for utilisation-dependent toll collection on interstate highways was implemented in July 2012, in the context of the "Moving Ahead for Progress in the 21st Century Act" (MAP-21). This law not only regulates the subsidy programmes for interstate infrastructure, but also comprises a fundamental reform of utilisation-dependent toll collection. While before the levying of congestion fees was only permitted in the context of pilot projects on certain sections of roads, from now on utilisation-dependent toll collection on interstate highways is regarded as the legally accepted best practice. In addition, implementation will be greatly facilitated by the legal obligation for all interstate highways receiving subsidies from the Federal Highway Trust Fund to ensure interoperability of all electronic toll collection systems by October 1, 2016.

3.1.7.4 Creative Solutions for Implementation Problems

Supporters of the status quo and opponents of the introduction of a utilisation-dependent road toll in Germany point to various implementation problems, in particular administrative costs, bypass behaviour, personal data protection concerns and public acceptance. It cannot be disputed that the costs of levying utilisation-dependent toll charges are higher than those of raising taxes or selling vignettes. However, the costs of toll collection systems have decreased hugely over the last decade (cf. Kossak, 2014, p. 291). In any case, the significant economic advantages of applying a congestion-based toll to the utilisation of scarce road infrastructure capacities are far greater than these administrative costs. Experiences with bypass traffic have been available since the introduction of the toll collection system for heavy goods vehicles in 2005; bypass behaviour was countered by extending the toll system to second-tier roads (for example dual carriageways). Fundamentally a driver is always confronted with the problem of weighing the higher costs of taking a detour against saving the toll charge. If realistic detour alternatives are available, the driver can choose between the toll charge and the additional time required for the detour. This freedom of choice does not constitute a fundamental argument against utilisation-dependent toll charges. Personal data protection was guaranteed when the vignette for heavy goods vehicles was introduced by mandating that the personal data collected are not permitted to be used for anything but billing purposes. Experiences in pilot projects in the USA show that when implementing a toll collection system, particular attention should be paid to a reliable guarantee of personal data protection, because this is of particular importance for the public acceptance of such systems (cf. Oregon Department of Transportation, 2007, Chap. 9). Acceptance can also be greatly increased by implementing transparent public participation in decision-making. In this context, experiences with public participation in infrastructure projects can be useful.

3.2 System Network Externalities (in Electricity Transmission Networks)

A special type of network externalities is to be found in electricity transmission networks. Because of certain laws of physics (Ohm's Law, Kirchhoff's Laws), electricity transmission is not confined to a specific path. As a result of the physical and technological characteristics of electricity transmission, system network externalities are of central importance in the electricity sector. Within an electricity network, it is not possible to transmit electricity between an entry and an exit point without simultaneously having an impact on the opportunity costs of network utilisation for the other network parts. Only in the special case of a network with two nodes is electricity transmission confined to this path. Consequently, system network externalities occur, when electricity transmissions impacts the opportunity costs of network utilisation for all network parts.

In contrast, congestion problems on railway tracks, airports or motorways constitute local (path-based) externality costs, because there is no systematic impact on any other infrastructure capacities. The electricity network scenario can only be transferred to transport networks in the extreme case of strictly complementary traffic flows on strictly complementary transport infrastructures. One example would be a hypothetical case of there being only two airports, A and B, in the relevant area of (derived) demand for take-off and landing slots, so that a departure from airport A would necessarily require a landing on Airport B (and vice versa).

3.2.1 Basic Characteristics of Electricity Transmission Networks

3.2.1.1 Loop Flows

The basic principle of loop flows can be illustrated by means of a simple stylised network with three nodes that are connected via three lines of equal length (cf. Hogan, 1992, p. 217; Keller, 2005, pp. 120ff.). The following Figs. 3.10 and

Fig. 3.10 Loop flows (entry point 1)

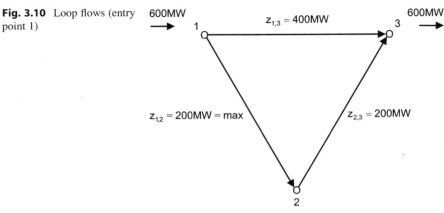

Fig. 3.11 Loop flows (entry points 1 and 2)

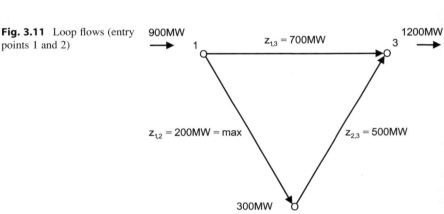

3.11 are based on such a stylised network. 1, 2, 3 designate the nodes and $z_{1,2}$, $z_{1,3}$, $z_{2,3}$ the electricity flow through the lines, with the capacity of line (1,2) limited to 200 MW. Flows disperse through the network in accordance with Kirchhoff's Laws: in every node the sums of incoming and outgoing electricity flows are equal (First Law), and in every electric circuit the sum of voltage drops due to the power extracted is equal to the voltage generated by the power fed in (Second Law). Thus electricity flows on the same path in opposite directions cancel each other out (netting principle). As the path from 1 to 3 via node 2 is twice as long as path (1,3), the flows will split at a ratio of 1 to 2. It is possible to generate 600 MW at entry point 1 and transmit it to exit point 3. Only 400 MW are transmitted via line (1,3); 200 MW are transmitted via line $(z_{1,2}+z_{1,3})$, which is twice as long (cf. Fig. 3.10).

Assuming demand in node 3 doubles to 1,200 MW, then even if generation capacity in node 1 were sufficient, it would not be technically feasible to feed 1,200 MW into node 1, although transmission capacity on line $z_{1,3}$ is unlimited

3.2 System Network Externalities (in Electricity Transmission Networks)

according to the assumptions made. The reason for this is the capacity constraint of 200 MW for line $z_{1,2}$ which would have to be exceeded, because according to Kirchhoff's Laws, $1{,}200 : 3 = 400 \text{ MW} > 200$ MW would have to be transmitted via line $z_{1,2}$, which is impossible. This problem can be solved by also feeding some electricity into node 2 (cf. Fig. 3.11).

The question arises how many MW must be fed into node 2 in order to guarantee that 1,200 MW will be available at node 3. Under the given assumption that electricity is only extracted at node 3, a parallel increase of the electricity feed-ins at nodes 1 and 2 will lead to the electricity flows in line $z_{1,2}$ cancelling each other out. Assuming that 900 MW are fed in at node 1 and 300 MW at node 2, the simultaneous feed-ins at nodes 1 and 2 result in opposite electricity flows in line $z_{1,2}$ between these nodes, and thus $1/3(900-300) = 200$, so that $z_{1,2 \, max}$ is not exceeded. At the same time, the sum of ingoing and outgoing electricity flows is the same at every node and the demand at node 3 is satisfied ($z_{1,3} = 700 = 2/3 \cdot 900 + 1/3 \cdot 300$ and $z_{2,3} = 1/3 \cdot 900 + 2/3 \cdot 300 = 500$).

Let $N = \{1, \ldots, n\}$ denote the numbers of nodes in the network. In the general case of $n > 3$ loop flows are also relevant. They result in an ever increasing likelihood of electricity flowing over several parallel paths. The case of $N = 2$ can be neglected, because in a network with only 2 nodes loop flows cannot occur. Nor will the case of $N = 1$ be examined, because in that case generation and consumption coincide (co-generation) and consequently no electricity transmission occurs at all.

3.2.1.2 Positive and Negative System Externalities

Let us again consider the 3-node network scenario illustrated in Figs. 3.10 and 3.11. The costs of electricity generation will at first be disregarded. In the following it is assumed that there is inelastic demand for extraction of a quantity of energy x in node 3, so that feed-in in nodes 1 and 2 has to be adapted in order to maintain the voltage balance. Let g_1 and g_2 denote the quantity of energy fed in node 1 and node 2, respectively.

Thus it must always be true that:

$$x = g_1 + g_2 \quad (3.49)$$

Although it is assumed that the lines between entry points 1 and 2, and exit point 3 have unlimited transmission capacity, feeding in at node 1 causes opportunity costs for electricity transmission. Theses occur on line $z_{1,2}$. The costs of electricity generation have to be separated from the costs of electricity transmission. Thus the problem to be solved is the minimisation of the transmission costs of the energy supplied at node 3.

Opportunity costs of feeding in at nodes 1 and 2 are denoted by $K(g_1)$ and $K(g_2)$.[26]

The Lagrange optimisation is:

$$L = K(g_1) + K(g_2) - \mu_{1,2}(z_{1,2} - z_{1,2max}) - \mu_{1,3}(z_{1,3} - z_{1,3max}) \\ - \mu_{2,3}(z_{2,3} - z_{2,3max}) \qquad (3.50)$$

It is assumed that the capacity limit is reached only on line $z_{1,2}$, whereas on lines $z_{1,3}$ and $z_{2,3}$ capacity is unlimited.

- Transmission Costs for Exclusive Feed-In in Node 1
  For the scenario illustrated in Fig. 3.10, exclusive feed-in in node 1 is physically possible. It is necessary to also use lines $z_{1,2}$ and $z_{2,3}$.
  It holds that:

$$\frac{\partial L}{\partial g_1} = \frac{\partial K}{\partial g_1} - \mu_{1,2}\frac{\partial z_{1,2}}{\partial g_1} - \mu_{1,3}\frac{\partial z_{1,3}}{\partial g_1} - \mu_{2,3}\frac{\partial z_{2,3}}{\partial g_1} = 0 \qquad (3.51)$$

Thus, for the welfare maximising network usage price it holds that:

$$p_1 = \mu_{1,2}\frac{\partial z_{1,2}}{\partial g_1} > 0; \mu_{1,2} > 0 \quad \text{due to the scarcity in line} \quad z_{1,2}. \qquad (3.52)$$

The price for the use of network capacity at node 3 is thus equal to the opportunity costs of using line $z_{1,2}$. From the perspective of the scarcity situation of lines $z_{1,3}$ and $z_{2,3}$ the two lines leading to node 3, the nodal price should be $p_1 = 0$, because there are no scarcities on these lines, and thus the shadow prices $\mu_{1,3}$ and $\mu_{2,3}$ are 0. However, these transmissions cause opportunity costs of network usage that accrue on line $z_{1,2}$.[27] This constitutes a network externality, as these opportunity costs cannot be assessed by considering one individual line $z_{1,3}$ or $z_{2,3}$ in isolation. This is also the crucial difference with regard to local externality costs which accrue on individual motorways or airports.

The conclusion to be drawn from this is that if there are system externalities, an isolated optimisation of the utilisation of individual lines does not reflect the actual opportunity costs of network utilisation. On the one hand, the technical feasibility of electricity transmission in the network might be compromised; in addition, the optimal utilisation of transmission capacity is not guaranteed. Even this very basic scenario (only 3 nodes, with only one load point, and at most two generation points) demonstrates the necessity of a central network coordinator, if network externalities exist.

---

[26] These are physical externalities which result immediately from the shadow price of a line's capacity constraint. The costs of electricity generation are not considered here.

[27] This is a corner solution, because scarcity occurs for an extraction of exactly 600 MW in node 3. With lower quantities there would be no scarcity.

3.2 System Network Externalities (in Electricity Transmission Networks)

- Transmission Costs for Feed-In in Nodes 1 and 2
  Here the starting point is Fig. 3.11. Conditions are assumed to be the same as in Fig. 3.10, except that double the amount of electricity is extracted in node 3. Extracting 1,200 MW in node 3 would be physically impossible, if feed-in occurred only in node 1. Because of Kirchhoff's Laws, additional feed-in in node 2 results in opposite electricity flows in line $z_{1,2}$.
  Thus: $\frac{\partial z_{1,2}}{\partial g_1} > 0$ and $\frac{\partial z_{1,2}}{\partial g_2} < 0$.

  Based on the Lagrange optimisation in Eq. (3.52), the welfare maximising network usage prices p_1 and p_2 in nodes 1 and 2 result as follows:

$$p_1 = \mu_{1,2} \cdot \frac{\partial z_{1,2}}{\partial g_1} > 0 \tag{3.53}$$

$$p_2 = \mu_{1,2} \cdot \frac{\partial z_{1,2}}{\partial g_2} < 0 \tag{3.54}$$

The opportunity costs of feed-in can be either positive or negative. Consequently, negative network usage prices are possible. In the 3-node example (cf. Fig. 3.11) this is shown as follows: Under equilibrium conditions the opportunity costs of network feed-in in nodes 1 and 2 correspond to the shadow price of the capacity constraint of line $z_{1,2}$. Based on the scenario chosen in this example, with incentives for generators in node 1 to feed in less electricity, and for generators in node 2 to feed in additional electricity (because of the electricity flows in opposite directions), in node 1 a positive network utilisation price results, whereas the generator in node 2 is compensated for its electricity feed-in at a price equal to the positive system externalities.

3.2.1.3 Reversal of Merit Order of Generating Plants Utilisation

Because electricity cannot be stored and demand for electricity is stochastic under real life conditions, it is efficient to use different types of generating plants with different marginal costs of generation. From this follows the so-called merit order of utilising generating plants with increasing marginal cost, differentiating between basic and peak load generating plants, with the latter having higher marginal costs. The plants with the lowest costs of generation will be used first, the one with the highest costs of generation will be hooked up last (cf. Crew & Kleindorfer, 1976).

Let λ_j denote production costs and β_j capacity costs of generating plant G_j. $j = 1, \ldots, J$, where $0 < \lambda_1 < \lambda_2 < \ldots \lambda_J, \beta_1 > \beta_2 > \ldots > \beta_J$.

If opportunity costs of network utilisation and network losses are neglected, optimal plant utilisation according to the merit order results, whereby the marginal generating plant feeds in at marginal costs $\lambda = \lambda_J$. λ denotes the system λ of electricity generation. It indicates the shadow price of generation, the marginal costs of the last unit or feed-in into the grid, in order to keep the balance between generation and demand. All generating plants with marginal costs $\lambda_j \leq \lambda$ contribute to the supply. If there are opportunity costs of network utilisation, they must be

Fig. 3.12 Reversal of merit order

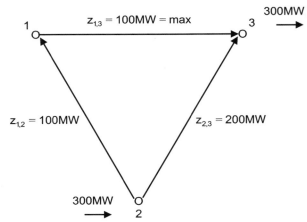

taken into consideration for feed-in decisions. It is assumed that at each entry point $j = 1, \ldots, J$ there is exactly one plant with marginal costs of generation λ_j.

The sequence of gross costs (generation and transport) is therefore decision-relevant, so that the merit order of the generating plants utilised which would result from the perspective of electricity generation without networks might well be reversed. In the trade-off between generation and transmission costs it may well be the generating plant with the higher costs of generation and the lower transmission costs that may prevail. If a generating plant has lower marginal costs of generation, but is situated at a disadvantageous location in the network where there are high opportunity costs for network usage, this plant might possibly not be able to feed in electricity. Even with a simple 3-line network it can be demonstrated that a scenario is possible where, due to a capacity constraint of the line between nodes 1 and 3, the generating plant at node 1 is forced off the grid, so that demand at node 3 can be met, and only the generating plant at node 2 with a less advantageous cost structure can feed in electricity.

Consider the 3-line network in Fig. 3.12: There are 2 generating plants, G_1 and G_2, located at nodes 1 and 2, respectively, meeting demand at node 3. Demand is 300 MW. The marginal costs of electricity generation are assumed to be constant. For generating plant 1 they are $\lambda_1 = v(G_1) = \frac{1}{2}v(G_2)$ and for generating plant 2 $\lambda_2 = v(G_2)$. If there is no scarcity in the network, G_1 will meet demand up to its capacity limit. Only at that point will G_2 be used. Now assume that there is a line capacity constraint between node 1 and node 3, whereby $z_1 = 100KW = \max$. Due to Kirchhoff's Laws in this scenario G_2 will have to produce the complete amount of electricity required.

The electricity flows are $z_{1,2} = z_{1,3} = 100KW$, $z_{2,3} = 200KW$. In contrast to the scenario illustrated in Fig. 3.11, where the electricity flows on line $z_{1,2}$ cancel each other out, here the flows add up on their path to load point 3. Each positive quantity of electricity generated in G_1 would result in the network being unable to fully satisfy the demand for 300 MW at node 3. The merit order is reversed, so that in this extreme case only the most expensive generating plant will feed electricity into the network (cf. Keller, 2005, pp. 183ff.).

The aim of the preceding paragraph was to examine the allocation problem posed by system externalities in a disaggregated manner. It was shown that, as long as a central network coordinator sets the prices for network usage for the individual nodes and thus internalises the economic system externalities, the decentrally operating power generators' feed-in decisions will inevitably be economically efficient. Generating plants that, due to a disadvantageous location, have to pay high feed-in charges while simultaneously having high generating costs will not be used.

This result can also be derived from the traditional integrated examination of electricity generation and transmission (cf. Bohn, Caramanis, & Schweppe, 1984; Schweppe, Caramanis, Tabors, & Bohn, 1988). Whereas in the disaggregated approach the central dispatcher sets network usage prices at each node—which generation and demand adapt to in their feed-in and extraction decisions—in integrated price-setting, one single price for generation and network usage is determined at each node.[28] From the perspective of the disaggregated approach of network economics (cf. Chap. 1) the exclusive optimisation of the transmission network is to be preferred, as it simultaneously enables competition on the level of electricity generation.

In this chapter disaggregated nodal prices have been derived for a transmission network of an exogenously given infrastructure. In Knieps (2013) disaggregated nodal pricing under variable infrastructure is derived simultaneously for the social welfare maximising benchmark as well as the profit maximising transmission network carrier. The optimal investment rule indicates that the infrastructure capacity of each line is extended to the point where the marginal costs of an extra unit of capacity are equal to the marginal benefit through reductions of system externalities on all lines. The optimal transmission pricing rule indicates the node-dependent injection and extraction prices raised by the network carrier. The generalised merit order indicates to the generators at which nodes injection is worthwhile so that generation costs and injection price do not exceed marginal willingness to pay on the wholesale market. Nodal prices at extraction nodes indicate to the consumers the sum of the (uniform) wholesale price plus the node-dependent extraction price. It is interesting to note that the investment rule under profit maximisation does not change compared to the socially optimal investment rule. Nevertheless, the profit maximising investment level taking into account profit maximising injection and extraction charges is smaller than the social optimal investment level taking into account socially optimal injection and extraction charges.[29]

---

[28] For a detailed analysis of the duality between the integrated and the disaggregated perspective, see Keller (2005, pp. 208ff.). It seems to be more than merely a question of the history of economic thought that the duality principle governing linear programming was developed in the context of electricity (cf. Dantzig, 2002; Kuhn, 2002).

[29] For the mathematical analysis see Knieps (2013).

3.2.2 Wind Energy and Efficient Electricity Transmission Networks

The gradual expansion of renewable energy and the successive phasing out of nuclear energy are currently at the center of controversial political debates. Well known issues are the valuation of the full costs and risks of alternative forms of energy, the extent of government subsidies for renewable energy and the intertemporal allocation of exhaustible resources. That injection of renewable energy can be expected to increase is beyond dispute. On August 4, 2010 the German Federal Government enacted a national action plan for renewable energy. For the electricity sector, the aim is to achieve an overall share of about 40 % for energy from renewable sources by 2020, with wind energy playing a progressively more important role.[30]

Less well known issues within the public energy policy debate are the new challenges for the electricity networks as a consequence of increasing generation of renewable energy. The growing injection of wind energy leads to an increasing pressure to increase node awareness of electricity transmission networks, taking into account the locational opportunity costs of injection and extraction of energy. In the past, generating plants were as much as possible built in the proximity of high demand areas, so that generating plants were distributed more or less evenly over the whole of Germany, with no definitive direction of electricity flows being discernible. However, as wind power plants are mainly situated in the north, the generation of wind energy leads to an increasing decoupling of areas of demand and areas of generation, and to growing electricity flows from the north to the south. Because short- as well as medium- and long-term forecasts of wind conditions are highly unreliable, electricity generation by means of wind power plants leads to high fluctuations in generation and limited possibilities of prognosis. Because of these progressively more complex allocation problems, decentralised locational decisions for future generating plants as well as short-term injection decisions, taking into account the scarcity signals of transmission network usage, become increasingly important.

The provisions of the German Energy Industry Act include various legal rules that limit entrepreneurial freedom in allocating electricity transmission capacities. This bundle of legal rules constitutes a coordinated set of interventions in the entrepreneurial freedom to design efficient allocation mechanisms for network capacities. According to § 15 (1) of the Electricity Network User Charge Ordinance[31] no access charges have to be paid for the injection of electrical energy. This means that the whole costs of network utilisation have to be borne by the extraction side. However, it is not only extraction, but also injection of electricity that has an

---

[30] Not only in Germany but also in the other European countries the increasing generation of renewable energy is a topical issue, see for example the Energy Roadmap 2050 of the European Commission (2011, p. 11).

[31] Verordnung über die Entgelte für den Zugang zu Elektrizitätsversorgungsnetzen (Stromnetzentgeltverordnung – StromNEV) vom 25. Juli 2005, Bundesgesetzblatt, Jahrgang 2005, Teil I Nr. 46, ausgegeben zu Bonn am 28. Juli 2005.

3.2 System Network Externalities (in Electricity Transmission Networks)

impact on the degree of network capacity utilisation. According to § 8 (1) of the Renewable Energy Sources Act[32] network providers are normally obliged to give priority to all electricity from renewable sources when injecting electricity into their networks. In addition, network providers are obligated to pay a fixed injection tariff to producers of renewable energy, based on average costs including cost of capital and independent of the current degree of network capacity utilisation. According to § 9 of the Renewable Energy Sources Act network providers must expand their networks if this is requested by energy producers in order to ensure the injection of renewable energy. However, from an economic perspective legally mandated network expansions are not efficient. Instead, the crucial factor is the interdependence of short-term decentralised allocation decisions on the basis of efficient network usage charges and long-term optimal network investment decisions.

As the marginal costs of generating wind energy are very low, wind energy would always be injected into the network as long as electricity transmission costs are neglected, even without legally mandated priority regulation. If transmission costs are taken into consideration, this may change, but only—given the low marginal costs of wind energy generation—if transmission costs are very significant. If the opportunity costs of network usage caused by wind energy are very high, the network provider must have the possibility to inject other forms of energy at more adequate locations. Providing scarcity signals of network usage the generalised merit order becomes relevant in such a way that the sum of generation costs and injection charges becomes relevant for the decentralised injection decisions of generators, irrespective of whether energy is renewable or non-renewable. However, if priority regulation is enacted, even inefficient network feed-ins of non-renewable energy to neutralise the wind energy surplus can become necessary in order to prevent voltage fluctuations leading to network instability.

Subsidising the capital costs of wind energy plants on the basis of the amount of wind energy injected into the network necessarily creates inefficient incentives to inject the maximum amount of wind energy, irrespective of the degree of network utilisation. If, however, the subsidy takes the form of lump sum payments, these inefficient incentives disappear. The necessity of priority injection for wind energy also disappears. In addition, optimum network usage charges based on the opportunity costs of network usage create the right economic incentives for investment in the expansion of electricity transmission networks, so that a regulatory obligation to expand the network becomes superfluous.

---

[32] Gesetz zur Neuregelung des Rechts der Erneuerbaren Energien im Strombereich und zur Änderung damit zusammenhängender Vorschriften, vom 25. Oktober 2008, Bundesgesetzblatt, Jahrgang 2008, Teil I Nr. 49, ausgegeben zu Bonn am 31. Oktober 2008 (Erneuerbare-Energien-Gesetz).

3.3 Questions

3-1: Hyper-Congestion
Explain the phenomenon of hyper-congestion and differentiate it from normal congestion.

3-2: Braess Paradox
What is the Braess paradox? Answer with the help of a graphic, taking into account under which preconditions there will be no incentives to use the intersecting road.

3-3: Loop Flows
Explain the basic principle of loop flows by means of a stylized network with three nodes that are connected via three lines of equal length.

3-4: System Externalities Versus Path-Based Externalities
In 1998, the contract path principle was established in the German electricity sector as the basis of pricing for network usage. According to this principle the network usage charge was based on the linear distance between an entry point and an exit point. Explain why this pricing structure could not be socially optimal.

References

Baumgarten, P., Huld, T., & Hartwig, K.-H. (2013). *Mautsystem für Fernstraßen in Europa.* Baden-Baden: Nomos.
Baumol, W. J., & Bradford, D. F. (1970). Optimal departures from marginal cost pricing. *American Economic Review, 60,* 265–283.
Bohn, R. E., Caramanis, M. C., & Schweppe, F. C. (1984). Optimal pricing in electrical networks over space and time. *Rand Journal of Economics, 15*(3), 360–376.
Braess, D. (1968). Über ein Paradox aus der Verkehrsplanung. *Unternehmensforschung, 258–268.*
Buchanan, J. M. (1969). External diseconomies, corrective taxes, and market structure. *American Economic Review, 59*(1), 174–177.
Büttler, H.-J. (1982). Grenzkostenpreise im Strassenverkehr. *Schweizerische Zeitschrift für Volkswirtschaft und Statistik, 2,* 185–203.
Button, K. (2004). The rationale for road pricing: Standard theory and latest advances. In G. Santos (Ed.), *Road pricing: Theory and evidence* (Research transportation economics, Vol. 9, pp. 3–26). Amsterdam: Elsevier.
Cohen, J. E., & Horowitz, P. (1991). Paradoxial behaviour of mechanical and electrical networks. *Nature, 352,* 699–701.
Crew, M. A., & Kleindorfer, P. R. (1976). Peak load pricing with diverse technology. *The Bell Journal of Economics, 7,* 2007–2231.
Dantzig, G. B. (2002). Linear programming. *Operations Research, 50*(1), 42–47.
Dewees, D. N. (1979). Estimating the time costs of highway congestion. *Econometrica, 47*(6), 1499–1512.
Edelson, N. M. (1971). Congestion tolls under monopoly. *American Economic Review, 61*(5), 873–882.
European Commission (1998). *Fair payment for infrastructure use: A phased approach to a common transport infrastructure charging framework in the EU.* White Paper COM(98) 466, July 1998.

References

European Commission (2011). *Energy Roadmap 2050*. Communication from the Commission to the European Parliament, the Council, the European Economic and Social Committee and the Committee of the Regions, Brussels, 15.12.2011, COM(2011) 885 final.

Hogan, W. W. (1992). Contract networks for electric power transmission. *Journal of Regulatory Economics, 4*, 211–242.

Keeler, T. E., & Small, K. (1977). Optimal peak-load pricing, investment and service levels on urban expressways. *Journal of Political Economy, 85*(1), 1–25.

Keller, K. (2005). *Netznutzungspreise in liberalisierten Elektrizitätsmärkten – Eine ökonomische Analyse der Entgelte für das Höchstspannungsnetz* (Freiburger Studien zur Netzökonomie 10th ed.). Baden-Baden: Nomos Verlag.

Knieps, G. (2006). Delimiting regulatory needs. In: *OECD/ECMT round table 129, transport services: The limits of (de)regulation* (pp. 7–31), Paris.

Knieps, G. (2013). Renewable energy, efficient electricity networks and sector specific market power regulation. In F. P. Sioshansi (Ed.), *Evolution of global electricity markets: New paradigms, new challenges, new approaches* (pp. 147–168). Amsterdam: Elsevier.

Knight, F. H. (1924). Some fallacies in the interpretation of social cost. *The Quarterly Journal of Economics, 38*, 582–606.

Kossak, A. (2014). Zur aktuellen Diskussion um eine Pkw-Maut in Deutschland. *Straßenverkehrstechnik, 5*, 287–294.

Kuhn, H. W. (2002). Being in the right place at the right time. *Operations Research, 50*(1), 132–134.

Lindsey, R. (2005). Recent developments and current policy issues in road pricing in the US and Canada. *European Transport/Trasporti Europei, 31*, 46–66.

Mohring, H. (1999). Congestion. In J. A. Gómez-Ibànez, W. B. Tye, & W. Clifford (Eds.), *Essays in Transportation Economics and Policy* (A Handbook in Honour of John R. Meyer, pp. 181–222). Washington, DC: Brookings Institution.

Mohring, H., & Harwitz, M. (1962). *Highway benefits: An analytical framework*. Evanston, IL: Northwestern University Press.

Morrison, S. A. (1987). The equity and efficiency of runway pricing. *Journal of Public Economics, 34*, 45–60.

Müller-Jentsch, D. (2013). Der Ring – smarte Citymaut in Stockholm – Auch der Nichtstau hat einen Preis. http://www.avenir-suisse.ch/author/daniel-mueller-jentsch/page/2/. May 30, 2014.

Oregon Department of Transportation. (2007). *Oregon's Mileage fee concept and road user fee pilot program*. Final Report, Oregon Department of Transportation, http://www.oregon.gov/ODOT/HWY/OIPP/ruftf.shtml. May 30, 2014.

Pigou, A. (1920). *The economics of welfare*. London: Macmillan.

Ramsey, F. (1927). A contribution to the theory of taxation. *The Economic Journal, 37*, 341–354.

Samuelson, P. A. (1992). Tragedy of the open road: Avoiding paradox by use of regulated public utilities that charge corrected Knightian tolls. *Journal of International and Comparative Economics, 1*, 3–12.

Santos, G. (Ed.). (2004). *Road pricing: Theory and evidence*. Amsterdam: Elsevier.

Schade, J. (2005). *Zur Akzeptanz von Travel Demand Management (TDM) Strategien – insbesondere Straßenbenutzungsgebühren*. http://vplno1.vkw.tu-dresden.de/psycho/projekte/afford/d_akzeptanz.html. May 30, 2014.

Schweppe, F. C., Caramanis, M. C., Tabors, R. D., & Bohn, R. E. (1988). *Spot pricing of electricity*. Boston: Kluwer.

Small, K. A., Winston, C., & Yan, J. (2005). Uncovering the distribution of Motorists' preferences for travel time and reliability. *Econometrica, 73*(4), 1367–1382.

Small, K. A., Winston, C., & Yan, J. (2006). *Differentiated road pricing, express lanes and carpools: Exploiting heterogeneous preferences in policy design*. AEI-Brookings Joint Center for Regulatory Studies, Working Paper 06-06, March, Forthcoming in Brookings-Wharton Papers on Urban Affairs.

Starkie, D. N. M. (1982). Road Indivisibilities, some observations. *Journal of Transport Economics and Policy, 16*(1), 259–266.

U.S. National Surface Transportation Infrastructure Financing Commission (2009). *Paying Our Way – A New Framework for Transportation Finance*. Final Report, February, http://financecommission.dot.gov. May 30, 2014.

Verhoef, E. T., & Small, K. A. (2004). Product differentiation on roads – Constraint congestion pricing with heterogeneous users. *Journal of Transport Economics and Policy, 38*(1), 127–156.

Wieland, B. (2005). The German HGV-toll. *European Transport/Trasporti Europei, 31*, 118–128.

Willig, R. D., & Baumol, W. (1987). Using competition as a guide: Railroad deregulation. *AEI Journal on Government and Society: Regulation, 1*, 28–35.

Winston, C. (1985). Conceptual developments in the economics of transportation - An interpretative survey. *Journal of Economic Literature, 23*(1), 57–94.

Wissenschaftlicher Beirat beim Bundesministerium für Verkehr, Bau- und Wohnungswesen (1999). Faire Preise für die Infrastrukturbenutzung – Ansätze für ein alternatives Konzept zum Weißbuch der Europäischen Kommission, Gutachten vom August 1999. *Internationales Verkehrswesen, 51/10*, 436–446.

Wissenschaftlicher Beirat beim Bundesministerium für Verkehr, Bau- und Wohnungswesen (2000). Straßeninfrastruktur: Wege zu marktkonformer Finanzierung, Empfehlungen vom Februar 2000. *Internationales Verkehrswesen, 52/5*, 186–190.

Wissenschaftlicher Beirat beim Bundesministerium für Verkehr, Bau- und Wohnungswesen (2005). Privatfinanzierung der Verkehrsinfrastruktur, Gutachten vom März 2005. *Internationales Verkehrswesen, 57/7+8*, 303–310.

Strategies for Price Differentiation

4.1 Basic Principles

A distinction has to be made between the application of price instruments on the markets for network services (retail markets) and price instruments on the markets for network infrastructure capacities (upstream markets). In the following it is assumed that free market entry is allowed on these markets and that the entrepreneurial search for price structures is not obstructed by regulatory interventions.

Price differentiation means that beside cost differences, demand aspects also have to be taken into consideration for pricing. Price differentiation can also occur in relation with product differentiation. Examples on the markets for network services are different quality classes in a train or airplane, different durations of contract (differentiation of contract period), quality of service differentiation in the provision of internet traffic according to different service classes. Examples on the markets for infrastructure capacities are different service classes for railroad track capacities, access to different highway lanes with high and low maximal congestion levels etc. In these cases it has to be noted that price differences which merely reflect the full costs of product differentiation (in particular the opportunity costs for quality of services differentiation) do not yet constitute price differentiation. The following definition takes these issues into account:

> ... price discrimination should be defined as implying that two varieties of a commodity are sold (by the same seller) [either to the same buyer or] to two buyers at different *net* prices, the net price being the price (paid by the buyer) corrected for the cost associated with the product differentiation (Phlips, 1983, p. 6).[1]

The term price differentiation and the term price discrimination are used synonymously. Within the traditional literature in the Anglo-Saxon world price discrimination is preferred. In order to avoid confusion with the legal concept of

---

[1] ["either to the same buyer or"] is added to the quotation in order to include quantity discounts.

discrimination and to underline its neutrality in terms of competition in the following the term price differentiation is used.

The phenomenon of price differentiation thus defined comprises a multitude of different price structures. Price differentiation includes any form of quantity discounts, peak load pricing, as well as differentiation according to the different willingness to pay of different consumer groups (student discounts etc.). The principle of Ramsey pricing taking into account the different price elasticities of various consumer groups also results in price differentiation, if the underlying marginal costs of production belong to the same product or to different product qualities rather than to multiproduct public utilities which may produce such different products as water, electricity and public transportation (cf. Sect. 3.1.4.2).

Three types of price differentiation can be distinguished (cf. Pigou, 1952, p. 279). First-degree price differentiation obtains when consumers' individual willingness to pay for different product units is fully exhausted. Second-degree price differentiation obtains when consumer groups' differing willingness to pay is skimmed by means of different prices, with all consumers within one group paying the same price. Through their purchase decisions, consumers self-select (endogenously) into a particular group. Third-degree price differentiation obtains when consumers can be divided into different consumer groups according to simple, objective criteria, with each group's willingness to pay being skimmed by means of different prices. In contrast to the endogenous market segmentation for second-degree price differentiation, with third-degree price differentiation consumers can be allocated to individual market segments exogenously.

The implementation of price differentiation strategies is based on two fundamental conditions:

- separability of the markets
- prevention of arbitrage

If a product can be bought on one market at a relatively low price and can then be sold on another market at a much higher price, price differentiation cannot be stable. Thus, the transferability of single units of a product between different markets must be difficult or impossible. Examples of such low potential for arbitrage are, for instance, the supply of individual households with gas, electricity or water, transport services, airport slots or railway tracks.

The limit of a more far-reaching price differentiation is reached, when the transaction costs for the pricing scheme become too high, that is, when the costs of avoiding arbitrage exceed the advantages of a more refined tariff system. However, this limit cannot be defined uniformly, but depends on the circumstances prevailing on each individual market. Further differentiation is no longer possible, if

- no further consumer group can be found that would buy more units, if prices were further reduced;

- no further market separation is possible (without arbitrage between individual groups occurring);
- cost recovery on the whole is endangered.

Price differentiation is significant on the markets for network services, as well as on the markets for network infrastructure capacities. If economies of scale are relevant, marginal cost prices cannot recover total costs. With the exception of perfect competition,[2] the instrument of price differentiation is relevant in all market forms, in a monopoly as well as in markets with functioning competition.

Insofar as price differentiation strategies, as compared to linear tariffs, are associated with an increase in market volume, buyers as well as suppliers of network services or network infrastructure capacities may benefit. Consequently, incentives for developing welfare enhancing pricing strategies emerge in a monopoly as well as under competition. The search for a suitable form of price differentiation constitutes an entrepreneurial task, because only the firms themselves, with the help of their market information, are able to implement adequate price differentiations. There is not one single optimal tariff scheme; instead, the limits of an additional price differentiation have to be investigated in the spirit of a search process.

4.1.1 Price Differentiation Through Peak Load Pricing

On markets with non-storable goods, systematic demand fluctuations over time (e.g. seasonal, or in the course of a day or week) lead to peak load pricing. A special characteristic of the peak load problem in network sectors is that there are differences in capacity utilisation both on the network services level and on the network infrastructure capacity level. The building of network capacities constitutes a joint product: capacity needed in peak load periods is also available in off-peak periods.

Applying peak load pricing based on demand estimates leads to time-dependent prices. Peak load pricing means that, during a given period, suppliers charge different prices for variants of a product that are identical with regard to location, quality and quantity but differ between times of consumption.

Network services are not necessarily produced by a single firm; instead, there may by different suppliers whose services build on each other in a complementary manner, with upstream network infrastructure capacities being supplied by the markets. Consequently, the structure of peak load tariffs is relevant not only on the markets for network services, but also on the upstream markets for network infrastructure capacities.

In the following the principles of peak load pricing are analysed for competitive markets as well as monopolistic markets. In both market forms firm peak case and

---

[2] Due to the large number of atomistic suppliers, economies of scale are excluded.

shifting peak case are differentiated. In the firm peak case, the utilisation of the capacity during the off-peak period constitutes a by-product, so that even at a price equal to operating costs (without contributing to the capacity costs) the capacity is not fully utilised during the off-peak period. In contrast, in the shifting peak case, capacity is fully utilised during the off-peak period, too.

4.1.1.1 Peak Load Pricing Under Competition

The standard model is based on the assumption that peak load demand and off-peak demand can be differentiated (cf. Steiner, 1957). In the meantime, generalisations for more than two demand periods have been developed (cf., e.g. Taylor, 1994).

The model developed by Steiner covers the problem of a socially optimal price structure maximising the sum of consumer and producer surplus. A differentiation has to be made between (constant) marginal capacity costs β and (constant) production costs b. However, because of the chosen model structure (in particular constant marginal costs of capacity) no positive profits occur (cf. Steiner, 1957, p. 587). Although Steiner did not explicitly develop the link to the market form of competition, his result characterises peak load pricing under competition (cf. Officer, 1966). The firms have identical production conditions. They take the prices of peak load and off-peak periods as given and supply the available capacity units. There is the possibility of both market entry and market exit, if production leads to profits or deficits, respectively. Indivisibility problems are disregarded because of the assumption that the marginal cost function for capacity is continuous.[3]

The socially optimal solution emerging under competition must factor in this characteristic of capacity provision being a joint product. D_c denotes demand for the non-storable services, resulting from a vertical addition of demand during peak load and off-peak periods. Let p_1 or p_2 denote the end customer prices in the peak load or off-peak period, respectively. In the firm peak case (cf. Fig. 4.1), the utilisation of the capacity during the off-peak period constitutes a by-product, so that even at a price $p_2 = b$ (that is, without contributing to the capacity costs) the capacity is not fully utilised during the off-peak period. The only crucial factor for capacity dimension is the demand during the peak load period, and peak load demand covers total capacity costs. Due to the non-existent scarcity during the off-peak period, however, this is not an instance of price differentiation. The difference between peak and off peak period prices only depends on the opportunity costs of capacity usage.

In contrast, in the shifting peak case, capacity is fully utilised during the off-peak period, too. The following Fig. 4.2 illustrates the shifting peak case (cf. Phlips, 1983, Fig. 8.3, p. 138).

The socially optimal total capacity x^o results in intersection point $D_c = 2b + \beta$. Output must be expanded in every period, until capacity is fully exhausted in every period (i.e. $x^o = x_1^o = x_2^o$); even during off-peak periods no capacities lie idle. The

---
[3] For an analysis of indivisibility problems, cf. Officer (1966, pp. 649ff.).

4.1 Basic Principles

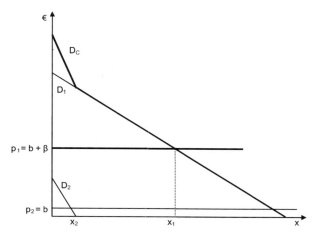

Fig. 4.1 Peak load pricing under competition (firm peak case)

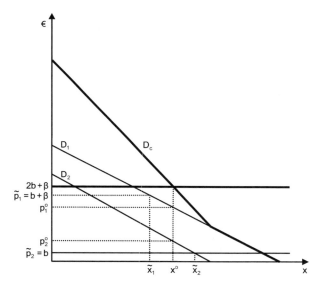

Fig. 4.2 Peak load pricing under competition (shifting peak case)

optimal prices associated with this scenario are p_1^o and p_2^o. This solution (the peak price and the off-peak price are contributing differently to capacity costs) is an economically desired form of price differentiation (cf. Steiner, 1957, p. 590). For peak load pricing in the shifting peak case optimal allocation is achieved, when the total opportunity costs for providing one additional capacity unit are exactly equal to the sum of willingnesses to pay of both demand groups (peak load and off-peak) for precisely this unit. The relevant opportunity costs thus do not refer to the specific utilisations of each demand group, but rather to the indivisible capacity unit as a whole, its opportunity costs incapable of being allocated to individual customers or demand groups. However, as in the optimum the willingness to pay, and thus the prices, for both demand groups are different, this is a form of price differentiation.

Price differentiation by means of peak load pricing is incentive compatible under competition. Assume that in the shifting peak case only peak load demand would contribute to the covering of capacity costs, so that $\tilde{p}_2 = b$ and $\tilde{p}_1 = b + \beta$. From this would follow that during the off-peak period there would be demand for quantities $\tilde{x}_2 > x^o$, and during the peak load period for $\tilde{x}_1 < x^o$. Only the capacity costs for $\tilde{x}_1$ units would be covered, but $\tilde{x}_2$ units would be needed. Because $\tilde{x}_2 > \tilde{x}_1$, peak load would shift to the off-peak period. Output during the off-peak period would then be larger than during the peak load period. However, the capacity costs would not be covered.

4.1.1.2 Peak Load Pricing in a Monopoly

The principle of peak load pricing for non-storable goods with demand fluctuating over time also holds in a monopoly scenario, so that a more constant utilisation of capacity can be achieved by means of temporal price differentiation (cf. Takayama, 1985, pp. 678–683). Based on the characteristic of capacity provision as a joint product, total demand for capacity D_c (total willingness to pay) again results from vertical addition of demand during peak load and off-peak periods. From this results the aggregated marginal revenue curve MR_c. Profit maximisation capacity results in intersection point $MR_c = 2b + \beta$ (shifting peak case) or $MR_c = b + \beta$ (firm peak case), respectively.

Just like under competition, the firm peak case as well as the shifting peak case can be differentiated. Due to marginalisation of the monopolist the capacity provided in the monopoly scenario is smaller than the capacity under competition.[4] The question arises whether under monopoly conditions, similar to the competitive situation, a firm peak case may also arise, resulting in excess capacity, even if off-peak does not contribute to capacity costs. Indeed, such a situation shall be shown in the following Fig. 4.3.

Due to the very low off-peak demand the relevant range of the vertical added demand D_c is equal to the peak load demand, thus the marginal revenue curve MR_c intersects $b + \beta$ at a profit maximising capacity at which only peak demand is served. Thus, capacity in the off-peak period is not scarce and its price is zero. Thus price differentiation does not occur.

In the shifting peak case (cf. Fig. 4.4), capacity is exhausted both during the peak load period and during the off-peak period. This leads to differentiated monopoly prices $p_1^m > p_2^m$, so that the following total monopoly rent results:

$$\pi^m = p_1^m x^m + p_2^m x^m - (\beta + 2b) x^m \qquad (4.1)$$

Thus for the monopolist it is also incentive compatible to apply price differentiation by means of peak load pricing on the basis of aggregated demand.

---

[4] This also holds for the general case of continuous time parameters (cf. Takayama, 1985, p. 683).

4.1 Basic Principles

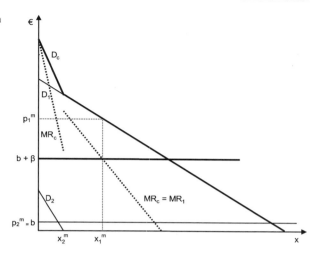

Fig. 4.3 Peak load pricing in a monopoly (firm peak case)

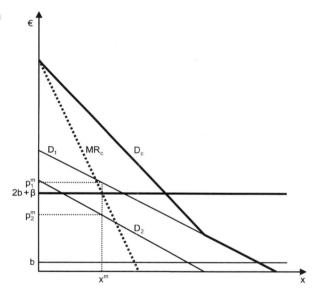

Fig. 4.4 Peak load pricing in a monopoly (shifting peak case)

4.1.2 Price Differentiation Through Optional Two-Part Tariffs

A two-part tariff consists of a fixed charge (base fee) plus a variable price component. If two-part tariffs are offered optionally beside one-part (linear) tariffs, the customer can choose either paying a higher variable user fee or paying a base fee plus a lower variable user fee. The principle of optionality is of fundamental importance to incentivise the self-selection properties of second-degree price differentiation. Starting from a linear tariff higher than marginal costs it is always possible to achieve a Pareto improvement by means of a price differentiation which

Fig. 4.5 Optional two-part tariffs

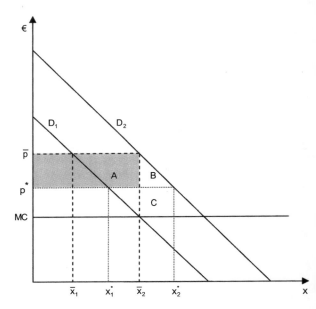

does not make matters worse for small consumers and improves matters for the firms and large consumers (cf. Willig, 1978). In order to not exclude small customers from network services, it is necessary to offer this two-part tariff optionally, so that small customers can receive the service at the original linear tariff $\bar{p}$. The basic principle of optional two-part tariffs can be illustrated by means of Fig. 4.5 (cf. Brown & Sibley, 1986, Fig. 4.6, p. 69).

Assume that there are two types of consumers: small consumers with demand $D_1(p)$ as well as large consumers with demand $D_2(p)$. Assume $D_1(p) < D_2(p)$ for all p. The two-part tariff (E, p^*) consists of a base fee E (shaded area) as well as a unit price p^*. This two-part tariff is offered optionally, in addition to a linear price $\bar{p}$ without an additional base fee, with $\bar{p} > p^* > MC$. Because $p^* < \bar{p}$ it follows that $x_2^* > \bar{x}_2$. E is chosen in such a way that the firm makes the same profit for the first $\bar{x}_2$ units of the large consumer as it does in a scenario where only the linear tariff is offered. This is achieved, by setting $E = \bar{x}_2 \cdot (\bar{p} - p^*)$. The large consumer has incentives to choose the two-part, optional tariff, because by transferring from $\bar{x}_2$ to x_2^* he can increase his consumer surplus by area B, and thus make an additional profit equal to area C. For the small consumer, on the other hand, paying the base fee is not profitable, because it would lead to a loss in consumer surplus equal to area A. Therefore the small consumer will continue to pay the linear price.

Quantity discounts (declining block rate tariffs) also constitute a form of price differentiation. These are non-linear prices, with the price per unit depending on the total quantity bought. Optional two-part tariffs are suitable for characterising and implementing non-linear price structures. There is a direct relation between non-linear multi-part tariffs with different tariff stages and a bundle of two-part tariffs from which the consumer can choose the two-part tariff that fits him best (cf. Brown & Sibley, 1986, pp. 80ff.).

The welfare improvements of optional two-part tariffs hold whenever economies of scale exist, independent of the underlying market form, for a cost-covering linear competitive price as well as for a linear monopoly price. In network sectors, due to the vertical complementarity between network infrastructure and network services, the question arises whether two-part tariffs for network infrastructure capacities can lead to distortions of competition on the markets for network services. This demonstrates once more how important it is that two-part tariffs are optional. Otherwise there would be the danger of small providers on the markets for network services being crowded out of the market, if paying a base fee was not profitable for them.

4.2 Price Differentiation in Network Sectors

4.2.1 Price Differentiation for Network Services

Price differentiation for network services can typically be observed in network sectors. Network services are non-storable, their provision is associated with relatively low marginal costs, but high fixed costs. In air transport, for example, the marginal costs of one additional passenger are only the costs for check-in, while the fixed costs comprise the opportunity costs for the airplane and the necessary flight personnel. The scenario is similar for bus and train transport. Examples for price differentiation are, for air transport, early bird rebates, frequent flyer programmes and yield management. For train transport, examples are rail passes, such as the Bahncard in Germany. On the markets for energy and telecommunications services, too, various forms of price differentiation are known (free terminal equipment, flat-rates, etc.).

An important indicator for the effects of the abolishment of legal barriers to entry on the airline markets is the increasing implementation of flexible price instruments (beside other marketing actions). Of particular importance are frequent flyer programmes (and related bonus systems) which constitute a special form of quantity discounts. Price differentiations between different customer groups and the associated different product characteristics (time of booking, flexibility in altering a booking, etc.) are quite common under competition. In the same way, the increasingly popular yield management systems that make price discounts dependent on the number of seats already sold are a permissible instrument under competition.[5]

---

[5] For a more detailed analysis, cf. Kahn (1993, pp. 396ff.), Locay and Rodriguez (1992), Borenstein and Rose (1994).

4.2.2 Price Differentiation for Railway Tracks

Tariff systems for the use of network infrastructures should be designed in such a way that they are able to simultaneously meet as fully as possible the criteria of being non-discriminatory, allocating scarce network infrastructure capacities efficiently, and fulfilling financing requirements. Traditional full cost allocation on the basis of administrative schemes for allocating the infrastructure costs to different user groups makes no economic sense and is known to be unable to solve the allocation and finance problem (cf. Sect. 2.2). However, pricing exclusively in accordance with (social) marginal costs does not necessarily meet all these criteria simultaneously either.

Traditionally the allocation of railway track capacities was determined by administrative measures on the part of the national railway monopolies. Examples of this were scheduling conferences, priority rules for different trains in case of delays, as well as discretionary individual measures by the train traffic management. Usage fees reflecting the scarcity of these track capacities were not levied, even if considerable scarcity of capacity occurred at particular times of the day or the year. Consequently, the decision at what time a section of railway track was used had no influence on pricing. Train companies therefore had no incentive to switch from peak times to off-peak times. Thus customer groups with high preference for punctuality and according willingness to pay had no possibility to use trains with guaranteed punctuality.

This problem can be solved by charging a (time-dependent) scarcity price for the use of railway tracks (cf. Knieps, 2006, pp. 217ff.). Train companies wishing to use a train on a highly frequented section of track would have to pay a track access charge reflecting the opportunity costs of capacity use (cf. Sect. 3.1). As a consequence, a freight train with traditionally low priority could be willing to pay more than an intercity passenger train with traditionally high priority, in order to prevent delay to particular production processes. As a consequence of efficient allocation of scarce track capacity track access charges are expected to be much higher in peak times than in off-peak times, resulting in peak load pricing on the (upstream) market for infrastructure capacities. Passing efficient track access charges through to the passengers of the train companies implies a higher train ticket tariff in peak times and a lower train ticket tariff in off-peak times, and thus creates incentives for the introduction of peak load pricing also on the (downstream) train service level.

Even on intensively used tracks optimal usage fees do not necessarily ensure full cost recovery. Economies of scale involved in the building of track infrastructures can be the reason why optimal access fees are insufficient to cover investment costs (cf. Sect. 3.1.4). Thus the issue of financing the deficit and, accompanying it, the issue of the ex ante politically prescribed level of cost coverage has to be dealt with. In order to make the incentives for achieving the necessary cost recovery a hard budget constraint for the infrastructure provider, the level of total cost coverage must not be left to ex post chance.

From June 1998 to April 2001 Deutsche Bahn AG offered an optional two-part track price system.[6] A customer had the option to either purchase a so-called InfraCard and pay a lower variable price per rail track capacity used, or pay a higher price for the actual track usage without buying the InfraCard. As the InfraCard constituted a base fee with a lower variable price per service unit (rail track) used, a general obligation to purchase the InfraCard would have barred smaller train companies from access to the track network of Deutsche Bahn AG altogether, or at least obstructed their access. Therefore it is very important to note that the purchase of the InfraCard was optional, and consumers were free to decide whether to buy it or not, depending on their demand characteristics. In the calculation of the variable price component of the InfraCard the rate of capacity utilisation and time table flexibility have been important components. Although the expected average capacity scarcity has been taken into account, no explicit peak load access charge system has been introduced.

The advantage of the optional two-part access charge system was that larger train companies had incentives to utilise the InfraCard as intensively as possible. Because in addition the smaller train companies, for whom the purchase of an InfraCard was not worthwhile, could also purchase track access, the implementation of this two-part system created incentives for more traffic on railway tracks.

4.2.3 Price Differentiation for Airport Slots

Traditionally, airport charges are basically dependent on the weight of the aircraft, their function being to help finance the airports, not to provide an efficient allocation of available capacities. Aircraft weight and flight distance are no indication of a flight's (marginal) contribution to the shortage of capacity available to airports, nor of the costs that ensue for all other airlines. Instead, the decisive factor in this respect is the demand at a particular time for airport capacities. In the short term, airport capacities are essentially unchangeable. In the event of unforeseen shortages, the typical solution is for airports to ration capacity on a first-come first-served basis.

Many airports reach the limits of their capacity during peak times. In the face of increasing congestion on airports, there are more and more appeals for the state to end this scarcity problem by expanding capacity further. It has to be kept in mind, however, that investments large enough to provide slots in abundance would, from an economic point of view, be a waste of resources (cf. Sect. 3.1.4). Instead investments should occur up to a point where the additional benefit of an extension of capacity is equal to the additional costs. Thus even with a level of investment that

---

[6] The introduction of track access charge systems was an immediate consequence of allowing competition on the tracks. Deutsche Bahn AG issued its first access pricing system on July 1, 1994 (cf. Knieps, 2005).

is optimal from an economic perspective, highly frequented airports will still have scarcity problems at peak times.

As soon as airport capacities are no longer available in abundance, the economic characteristics of airport capacities regarding take-off and landing slots transform from having the character of a public good into a private good (cf. Knieps, 2006, pp. 13ff.). As a consequence the specification and definition of what has become scarce and at what time the scarcity developed is called for. Even when trading more ordinary goods, for example grains, the micro-economic problem of defining categories of goods (e.g. types of grains) becomes relevant; however, the precise point in time of the transaction is usually not of crucial importance. Matters are very different for airport capacities; here a multitude of resources have to be synchronised as precisely as possible. The definition of a slot already opens up a considerable range of alternative options, which may have a crucial impact on the transaction potentials. If the allocation of a take-off slot only means that an airline has the right to take off within a relatively large time interval, this right is for some airlines of considerable less worth than the guarantee to be permitted to take off at a very specific point in time, without any delays. In contrast, other airlines may prefer more flexible operating times. Therefore slot trading presupposes a definition of slots that takes into account both the requirements of airlines (and their passengers), and the operational strategies of the airport operator.

The slot charges should vary according to the degree of capacity utilisation during a day and depending on the season, as capacity utilisation for the same flight may vary. This would enable peak time take-off and landing rights to be allocated more efficiently. This type of slot charge would operate as peak load pricing of (upstream) infrastructure capacities.[7] Another advantage of levying time-dependent scarcity-based slot charges allowing efficient short-term allocation of slots is that when congestion fees are high during peak periods there is no incentive to hoard slots.

Within the European Community a common framework regulation on airport charges applied to airports located in the Community that are above a minimum size has been introduced by Directive 2009/12/EC.[8] These regulations do not prescribe a specific allocation mechanism, as long as the aim is to cover the decision-relevant costs of the airport. Airport management is free to apply efficient airport charges taking into account capacity scarcity of slots (cf. Knieps, 2013).

To date only a few airports levy capacity shortage-based take-off and landing fees, with Great Britain being the forerunner in Europe. At London's Heathrow and Gatwick airports landing fees based on peak load pricing have been charged since the early 1970s (cf. Civil Aviation Authority, 2001 Annex). In the morning and

---

[7] A (non-time based) congestion charge would still have to be levied, even if there was no change in capacity utilisation over time and no fluctuation in the level of the airport usage. For the detailed analysis of congestion charges see Chap. 3.

[8] Directive 2009/12/EC of the European Parliament and of the Council of 11 March 2009 on airport charges, 14.3.2009, OJ L 70/11.

evening hours a standard peak landing fee applies regardless of aircraft weight. The principle of peak load pricing is also applied in relation to aircraft passenger and parking fees. As peak load times for aircraft landing, passenger clearance, and parking of airplanes are not the same, different peak periods are defined for each of these services, with different peak load prices. Strategies to deter smaller aircraft from using the airport at peak times by introducing base fees or minimum landing fees can also be observed (cf. Knieps, 2006, p. 24).

4.3 Questions

4-1: Peak Load Pricing
Determine the optimal capacity under peak load pricing in the shifting peak case under competition. Give reasons why this case constitutes price differentiation.

4-2: Two-Part Tariffs
Explain the significance of the optionality of two-part tariffs.

4-3: Market Form and Price Differentiation
What influence does the market form have on price differentiation?

References

Borenstein, S., & Rose, N. L. (1994). Competition and price dispersion in the U.S. airline industry. *Journal of Political Economy, 102*(4), 653–683.
Brown, S. J., & Sibley, D. S. (1986). *The theory of public utility pricing*. Cambridge: Cambridge University Press.
Civil Aviation Authority. (2001). *Peak pricing and economic regulation – Annex*, London.
Kahn, A. E. (1993). The competitive consequences of hub dominance: A case study. *Review of Industrial Organization, 8*, 381–405.
Knieps, G. (2005). Railway (De-)regulation in Germany. *CESifo DICE Report, 3/4*, 21–25.
Knieps, G. (2006). *Delimiting regulatory needs, Round Table 129, Transport Services: The limits of (De)regulation*, Paris, pp. 7–31.
Knieps, G. (2013). *Market versus state in building the aviation value chain*. Institute for Transport economics and regional policy discussion paper no. 146, Freiburg University.
Locay, L., & Rodriguez, A. (1992). Price discrimination in competitive markets. *Journal of Political Economy, 100*, 954–965.
Officer, L. H. (1966). The optimality of pure competition in the capacity problem. *Quarterly Journal of Economics, 80*(4), 647–651.
Phlips, L. (1983). *The economics of price discrimination*. Cambridge: Cambridge University Press.
Pigou, A. C. (1952). *The economics of welfare* (4th ed.). London: Macmillan.
Steiner, P. O. (1957). Peak loads and efficient pricing. *Quarterly Journal of Economics, 71*, 585–610.
Takayama, A. (1985). *Mathematical economics*. Cambridge: Cambridge University Press.
Taylor, L. (1994). *Telecommunications demand in theory and practice*. Dordrecht: Kluwer.
Willig, R. D. (1978). Pareto superior nonlinear outlay schedules. *Bell Journal of Economics, 9*, 56–69.

Auctions

5.1 Basic Principles

An auction is a market mechanism which, by means of explicit rules, transfers market players' bids into an allocation of resources. An auction yields an unequivocal result as to which market player gets a specific object at which price (e.g. McAfee & McMillan, 1987, p. 701). For a long time, auctions have played a significant role in many sectors of the economy. Auctions of paintings, antiques, wine, or cattle are just some relevant examples. Due to the success of the Internet, the auction as a market place for the exchange of goods has gained yet more importance. In network sectors auctions also have considerable potential as allocation mechanisms for network services (e.g. providing bus services) and for the provision of network infrastructure capacities (e.g. take-off and landing slots on airports). In this context, auctions of public resources, for instance radio frequencies, on the basis of which network infrastructures can be built, are the best known examples. Auctions are by no means the only mechanism for a market-compatible allocation of scarce goods; the usual market transactions at list prices and even barter trade (with and without side payments) are also common.

On markets the owner of a private good is usually free to decide for himself how to utilise his property and what allocation mechanism to use, if he wants to sell it. The comparative advantages of different allocation mechanisms thus emerge endogenously in the competitive process. Matters are different when a public authority is involved; here transparency and non-discriminatory conditions are of prime importance, no matter if the public authority functions as supplier or customer. Thus the awarding of public contracts is typically associated with an obligation to tender. For the commissioning of services of general economic interest which are subject to public service obligations, bidding procedures for the right to provide the unprofitable service at minimal subsidies are relevant. For the allocation of scarce infrastructure capacities (e.g. airport slots) as well as for awarding rights of way, frequencies etc., auctions are also particularly suitable.

Although auctions have been common in practice for a long time, auction theory has only been developed since the middle of the last century, as a branch of non-cooperative game theory. Non-cooperative behaviour can be specified with the help of the Nash behavioural assumption (Nash, 1951). Applied to auctions, this means that the rival bids are assumed as given and incapable of being influenced. In this context each bidder tries to make the decision that is most advantageous for him. A group of strategies is termed a Nash equilibrium if—assuming that all other bidders' strategies are constant—no bidder can achieve a higher benefit by choosing another strategy.

Auction theory emphasises the analysis and development of different types of auctions. Model analyses have so far taken considerable less interest in determining the object of the auction and examining the institutional framework of auctions. The fundamentals of auction theory are by now part of the standard repertoire of introductory microeconomics courses (cf. e.g. Varian, 2010, Chap. 17).[1]

In the literature on auction theory different criteria for evaluating the "quality" of an auction design are used, in particular achievable revenue, incentives for collusion, the possibility to take into account value interdependencies between auction objects, and the degree of complexity of implementation (cf. Klemperer, 2002; Milgrom, 1987; Robinson, 1995; Wolfstetter, 1996).

The conclusion to be drawn from the substantial literature is that no single ideal auction design exists that could be universally recommended: "Furthermore, anyone setting up an auction would be foolish to follow past successful designs blindly; auction design is *not* 'one size fits all'" (Klemperer, 2002, p. 187). Thus the aim is not the construction of an "ideal" auction mechanism; instead, in the spirit of comparative institutional analysis (Demsetz, 1969), the main objectives must be to examine the role of auctions in the context of alternative allocation mechanisms and to work out the determining factors for institutional competition in the quest for better auction mechanisms.

This chapter will focus on the potentials of invitations to tender and auctions in network sectors and on the resulting network-specific problems, which auction theory has so far analyzed inadequately or not at all. It will be shown that in particular the different objectives of invitations to tender, as well as a suitable definition of the auction object, are of special importance.

5.1.1 Elements of Auction Design

The economic theory of auctions distinguishes between two polar cases for the distribution of the bidders' evaluations of the object being auctioned and two basic auction formats based on the distinction between open versus sealed bids auctions (e.g. Lipczynski, Wilson, & Goddard, 2005, pp. 389ff.).

---

[1] A more detailed overview can be found e.g. in Klemperer (2004, pp. 9–72) and in Lipczynski et al. (2005, Chap. 11).

5.1.1.1 Private-Value Auctions Versus Common-Value Auctions

Asymmetrical information is a central element in auctions. The seller has no perfect information on the bidders' individual willingness to pay for the object of auction, and the bidders have no perfect information on the other bidders' valuations or willingness to pay.

- Independent Private-Value Auctions
  Each bidder evaluates the object individually, so that the private value corresponds to the individual willingness to pay. These individual evaluations differ from one bidder to the next, and there is no objective reference value (inherent value), which can be determined by the bidders. An example for this is the auction of a painting by an unknown artist, which is evaluated differently by the individual bidders.
- Pure Common-Value Auctions
  The object has a definitive, unequivocal value which is the same for all bidders. However, this value is not known to the bidders. On the basis of private information varying between bidders, each bidder forms an independent assessment of this objective value. An example for this is an auction for drilling rights for oil, where the amount of crude oil available in the oil field determines the objective value of the drilling rights, which—as long as the oil price is given—is assessed at the same value by all bidders. At the time of the auction, however, the bidders do not know the actual amount of crude oil, although every bidder has made a private assessment. A bidder's private assessment can be influenced by his knowledge of the other bidders' assessments.
- Hybrid Forms of Private-Value and Common-Value Auctions
  The individual valuation of an object depends both on individual taste and on the objective reference value (for example, the expected market price). Examples for this are collectors' items or antiques, where the market-compatible estimated price may be either overbid or underbid.

5.1.1.2 Open Versus Sealed-Bid Auctions

A distinction has to be made between open auctions, where bidders know the other bidders' offers and can make counter offers (English and Dutch auction), and sealed-bid auctions, where bidders make their offers simultaneously, without making them known to the other bidders (first-price sealed-bid auction and Vickrey auction). In addition, there are a large number of hybrid forms, such as for instance the English-Dutch auction.

- English Auction
  This is an ascending-bid, open auction. First, a reservation price (a minimum price that would be acceptable to many bidders) is set, which is then successively raised, until only one bidder remains. The last remaining bidder who offered the highest bid gets the object and the auction ends.

- Dutch Auction
  This is a descending-bid, open auction. First an initial price is called, at a level so high that no bidder would be prepared to pay it; then the price is successively lowered, until one bidder is willing to pay the current price. This bidder is awarded the object at the price he bid and the auction ends.[2]
- First-Price Sealed-Bid Auction
  Here every potential buyer submits one independent individual sealed bid, without knowing what the other bids are. The highest bidder is awarded the object at the price he bid. First-price sealed-bid auctions are known for example to be used by public authorities when auctioning extraction rights for oil, natural gas, minerals, etc.[3]
- Vickrey Auction (Second-Price Sealed-Bid Auction)
  Every bidder submits one independent individual sealed bid, without knowing what the other bids are. The highest bidder is awarded the object, but at a price equal to the second-highest bid.[4] Vickrey auctions are used for auctioning stamps and for some auctions on the Internet. This type of auction was also under debate for the auctioning of electricity (cf. Lucking-Reiley, 2000). In spite of their theoretical elegance, however, Vickrey auctions are used comparatively seldom (cf. Rothkopf, Teisberg, & Kahn, 1990).
- English-Dutch Auction
  This is an ascending-bid auction, where the price increases, until only a small, predefined number of bidders remain. The remaining bidders are then asked to submit a sealed bid that cannot be lower than the last open bid. The winner pays a price equal to the highest bid. Auctions on eBay have certain similarities to English-Dutch auctions. Since eBay specifies the end of the auction, that is, the point in time up to which bids can be made, the final phase has the characteristics of a sealed-bid auction. Because of the specified deadline many bidders only bid during the last seconds, so that they cannot react to other bidders' offers any longer (cf. Klemperer, 2002, pp. 181f.).

5.1.1.3 Auctions of Individual Objects Versus Auctions of Several Units of an Auction Object

A large part of auction theory examines auctions of individual, non-divisible objects. The simultaneous auction of several units of an auction object has been examined to a much smaller extent in the literature so far, but constitutes an increasingly active field of research (cf. Klemperer, 2004, p. 29). The units can be either homogenous or differentiated for quality. The following cases can be distinguished:

---

[2] This type of auction is called Dutch auction because it became famous in the Netherlands for auctions of tulip bulbs.

[3] A particular variant of the first-price sealed-bid auction, in which a successful bidder might, in the context of an invitation to tender, under certain conditions be permitted to withdraw his bid is called Swiss auction (cf. von Ungern-Sternberg, 1991).

[4] The term Vickrey Auktion goes back to William Vickrey, who wrote essential contributions to auction theory (e.g. Vickrey, 1961).

- Each bidder only wants to buy at most one unit of the auction object (e.g. non-transferable tickets).
- The bidders can choose to bid for individual units, or, alternatively, for combinations of units (e.g. licenses for bus transport). These are combinatorial auctions, and their advantage is that bidders can formulate their preferences in more detail. This is particularly the case if the different units are complementary (cf. Cramton, Shoham, & Steinberg, 2006, p. 4).

5.1.2 Fundamental Problems of Auction Theory

5.1.2.1 Common-Value Auctions and Winner's Curse
Winner's curse is a term for the danger that the winner of an auction with the highest individual assessment will overestimate the objective value of the auction object. The winner of the auction is then actually the loser, because he pays more than the real value of the object of auction. In contrast, winner's curse does not occur in private-value auctions, because here every bidder evaluates the auction object according to his subjective valuation and related willingness to pay.

5.1.2.2 Incentive Compatible Versus Strategic Bidding[5]
In an auction, incentive compatibility means that bidders make their bids in accordance with their individual willingness to pay. In contrast, strategic bidding means that it may be profitable for bidders to make bids lower than their individual willingness to pay. In the following it will be shown which auctions are incentive-compatible and which elicit strategic bidding. In the case of strategic bidding it is assumed that bidders are risk neutral and exclusively focused on the expected value of their auction results. They are indifferent between an assured acceptance of their bid with rent a and a 0.5 probability of winning the auction with rent $2a$.

- Incentive Compatibility of the English Auction and the Vickrey Auction
  In the English auction the optimal bidding strategy is to bid until the price reaches one's individual willingness to pay and withdraw from the auction as soon as the price becomes higher. The bidder with the highest willingness to pay wins the auction. If the bid is successful, the bidder gets a consumer surplus equal to the difference between his individual willingness to pay and the successful bid. If the bid does not succeed, the bidder wins or loses nothing.

  In the Vickrey auction the bidder with the highest bid pays a price equivalent to the second-highest bid. As in the English auction, the optimal bidding strategy is to make a bid equivalent to one's individual willingness to pay; because if a bidder makes a bid that is higher than his individual willingness to pay, this may result in his winning the auction and having to pay the equivalent of the second-

---

[5] For a more detailed and mathematical analysis the reader is referred to Lipczynski et al. (2005, pp. 389ff. and 704ff.).

highest bid, which may still be higher than his individual willingness to pay.[6] If a bidder makes a bid that is lower than his individual willingness to pay, in the worst case this may result in his not winning the auction, even though his individual willingness to pay is not exhausted.[7]

- Strategic Bidding in the Dutch Auction and the First-Price Sealed-Bid Auction
In the Dutch auction the optimal bidding strategy is for a bidder to wait until the price has fallen under his individual willingness to pay, and then to bid. If his bid is accepted, he earns a bidder's surplus equal to the difference between the bid and his willingness to pay. If another bidder wins the auction, he suffers no loss, due to the assumed risk neutrality. In contrast to the English auction, risk aversion impacts the optimal bidding strategy. A risk averse bidder is prepared to accept a reduction of his remaining consumer surplus (in case he wins the auction) in exchange for a higher probability of winning the auction (cf. Lipczynski et al., 2005, p. 403).

In the first-price sealed-bid auction it is also profitable for the bidder to make bids that are lower than his individual willingness to pay. It is a trade-off between the increase of the expected bidder's surplus because of a lower bid, and the decrease in the probability of winning the auction. Assuming equally distributed, independent individual willingnesses to pay V_i of bidders i ($i = 1, \ldots, N$) and risk neutrality, it can be shown that in a Nash equilibrium the optimal bidding strategy for each bidder consist of making a bid B_i as follows:

$$B_i^* = \frac{N-1}{N} \cdot V_i, \quad i = 1, \ldots, N \quad (5.1)$$

If there are only two bidders, each bidder makes a bid equal to half of his individual willingness to pay. If there are three bidders, each bidder makes a bid equal to $2/3$ of his individual willingness to pay. In case of 100 bidders, each bidder makes a bid equal to $99/100$ of his individual willingness to pay. Thus it can be seen that with an increase in the number of bidders the optimal bid approaches the individual willingness to pay. Risk averse bidding behaviour also has an influence on optimal bidding strategy, because risk averse bidders make higher bids than risk neutral bidders (cf. Lipczynski et al., 2005, pp. 394ff.).

5.1.2.3 Price and Quality Differentiation in Auctions of Several Units of an Auction Object

An essential characteristic of auctions, as compared to other allocation mechanisms, is that they disclose bidders' individual willingness to pay. When several units of an object are auctioned, bidders' different individual willingnesses to pay can be exhausted. Under the assumption that each bidder either buys one unit

---

[6] In the most favourable case the amount of the bid higher than his individual willingness to pay is irrelevant, because the second highest bid is not impacted by it.

[7] In the most favourable case the amount of the bid lower than his individual willingness to pay is irrelevant, because the second highest bid is not impacted by it.

at auction or none at all, the first unit will go to the bidder with the highest willingness to pay and the last unit to the bidder with the lowest willingness to pay.[8]

The different units of the auction object may also differ in quality. When such units are auctioned, the different product qualities are reflected endogenously in the prices. This has to be distinguished from an auction of objects that belong to different product categories. The latter type of auction neither constitutes price differentiation nor an auction of several units of an auction object.

5.2 Auctions in Network Industries

5.2.1 Network-Specific Particularities

In auction theory the decision to use an auction as an allocation mechanism in the first place is typically already assumed. In addition, for reasons of simplicity it is often assumed that the object of the auction is well-defined and the objective of the auction is assumed to be the maximisation of revenue. The focus is then on the efficiency of the auction design.

In order to apply auction theory to network sectors it is necessary to distinguish between the following problem areas:

- Obligations for Public Authorities
  The starting point is to question what freedom of choice the seller has in choosing and designing the allocation mechanism. In a market economy, the reference point is the potential buyer's (bidder's) and the seller's voluntary decision whether they want to participate in an auction or not. The seller can choose between existing auction platforms or build his own auction platform. For contracts (beyond a certain minimum amount) that are awarded by a public authority, however, an invitation to tender is compulsory. Although public resources (e.g. licenses) do not necessarily have to be allocated by means of an auction, auctions are particularly suitable for ensuring mandatory non-discrimination and transparency. Invitations to tender are also important for commissioning politically desired loss-making universal services. As loss-making universal services are not supplied spontaneously under competition, the type and extent of these services have to be determined in the political process and their financing must be guaranteed at the same time. In doing so, a clear-cut distinction between the commissioning function on the one hand and the supplier of network services (contractor) on the other hand is necessary.[9] Because invitations to tender and auctions are market-compatible allocation mechanisms on opened markets, they are on principle not compatible with grandfathering. Auctions can only be used as allocation mechanisms for the supply of network

---

[8] For the concept of price differentiation see Chap. 4.

[9] The role of universal services in network industries is discussed in Chap. 7.

services, if the necessary institutional precondition of a comprehensive market opening is fulfilled (e.g. a market for slot allocation on airports has been introduced etc.).
- Objectives of Invitations to Tender and Auctions
  Invitations to tender and auctions, like other market-based allocation mechanisms, aim to balance sellers' and buyers' interests as far as possible.

The objective of allocative efficiency requires that the object of a public tender goes to the bidder with the highest valuation as long as the price is higher than the seller's reservation price. However, a scenario where the bidder overpays the object of a public tender (winner's curse) should also be avoided. The same holds for contracts awarded by public authorities in the form of a minimisation of public expenditures.

Auction theory is focused on the seller's revenue objective. Collusion between bidders is to be prevented as far as possible; discrimination to the disadvantage of new bidders must also be avoided. This is the only way to achieve the maximum quasi-rent for an auction object. However, in network sectors the objective of revenue maximisation can lead to a monopoly rent that is undesirable from an economic perspective. Designing auctions in network sectors in such a way that sellers realise the highest expected profit can thus come into conflict with the objective of price level regulation.[10]

5.2.1.1 Auctions and Network Competition

In network sectors where a single provider is able to serve the relevant market at lower costs than several providers (natural monopoly), auctions serve a special function. Auctioning off the right to supply a market that is a natural monopoly can, under certain conditions, replace the lack of active competition in the market. According to Demsetz (1968, p. 58) the following two conditions are crucial for the functioning of such an auctioning process:

- competition on the input markets (many potential bidders), as well as
- no collusion between competing bidders.

Demsetz offers no concrete suggestions on the institutional design of the tendering procedure. His main objective is to show the disciplinary effects of potential competition in case of a natural monopoly.

So far, the potentials of invitations to tender and auctions in network sectors have not been fully exploited. A fundamental reason for this is that the obligation to implement market-compatible allocation procedures is not consistently enforced, and subsequently invitations to tender and auctions are not undertaken.

In the transportation sector, for instance, grandfathering and closing up markets by means of incumbent privileges are still of considerable importance. Active and potential competition in public transport can only be effective, if exclusive rights

---

[10] The role of market power regulation in network industries is analyzed in Chap. 8.

and incumbent privileges are abolished. Competition for a market to provide network services is only possible, if concessions for serving individual lines or a bundle of lines are auctioned to the most efficient bidder. The aim should be to determine those firms that are ready to provide services—predefined in quality and quantity—at the most favourable terms and conditions. Concessions should be granted for a limited period of time and without preference provisions or grandfathering.

Grandfathering still plays an important role for slot allocation on airports. Take-off and landing slots in Europe are still not redistributed in accordance with changing demand, but remain in the hands of the airline to which they were initially allocated, even if that airline does not use them or if another airline would have a higher valued usage. With an auctioning of take-off and landing rights the advantages of long-term flight scheduling could be maintained, but at the same time the market would be opened up to newcomers.[11]

5.2.1.2 Complexity of the Auction Object in Networks

No matter which allocation mechanism is used, the transaction object must be defined first. This issue is hardly examined in auction theory, because auction objects can usually be traded individually. Matters are different, however, if network interdependencies exist. A basic distinction has to be made between the smallest economically meaningful auction object on the one hand, and the externalities between different auction objects on the other. The more comprehensive the auction object chosen, the smaller the danger of negative externalities and the potentials for value interdependencies between the objects. Vice versa, the smaller the auction object becomes, the bigger the danger of negative externalities.[12]

Interdependencies and network externalities vary considerably on the different network levels. A distinction has to be made between network externalities in the provision of network services and network externalities in the provision of network infrastructure capacities. Whereas the smallest auction objects can be the reservation of a train track, train control systems typically have large-scale dimensions. In order to provide network services, complementary network capacities may be required. Thus for example a flight requires not only a take-off, but also a landing right. However, the quantities of take-off rights and landing rights on different airports are independent of each other, as they are used for many different destinations, and can therefore also be traded independent of each other.

---

[11] For the role of auctioning of take-off and landing slots in the liberalisation of the aviation value chain see Knieps (2006, p. 22).

[12] For the problem of negative externalities due to interferences between adjacent frequencies and the resulting need to delimit the rights of operators to transmit signals which are interfering or might interfere as precondition for market transactions resulting in an optimal utilisation of rights see Coase (1959, pp. 25ff.).

5.3 Disaggregated Invitations to Tender and Auctions in Network Sectors

The manifold potentials of invitations to tender and auctions are important, on the level of network services as well as on the levels of infrastructure management, network infrastructure and public resources, on the basis of which network infrastructures can be built. In the following, individual case studies are chosen as illustrations.

5.3.1 Invitations to Tender in Public Transport

In public transport there are requirements for integrated tariff systems and time table coordination; however, these requirements do not make it necessary for the public authority inviting tenders to also undertake the entrepreneurial tasks. Specifying the parameters of service quality at a politically desired tariff by no means implies the simultaneous taking on of entrepreneurial functions. Instead, the latter should be contracted out in the context of a competitive invitation to tender, simultaneously determining the most cost-efficient transport supplier.

Competition for a market is made possible, if concessions for serving individual lines or a bundle of lines are auctioned off to the cheapest supplier. The aim should be to determine which firms are prepared to provide minimum service—predefined in quality and quantity—at the most favourable terms and conditions. Competition in a market is made possible, if—for instance in profitable long-distance line traffic—several suppliers can obtain a concession simultaneously, or if quantitative market access restrictions disappear completely and tariffs can freely develop. It is to be expected that competitive tendering will be particularly successful in public transport, because both buses and trains are mobile and not tied to a specific line or a specific geographical network; consequently, there are no irreversible costs on which strategic behaviour could be based. The precondition for this is, however, that no supplier of bus or train transport is granted preferential access to the infrastructure or preferential treatment in time table scheduling. Consequently, track price catalogues must not contain discriminatory elements at the expense of market entrants.

When tendering these loss-making network services, the best bid is the one providing these services with the lowest subsidy requirements. The strategy of tendering a bundle of both lucrative and loss-making network services, with the objective of lowering subsidy requirements—accompanied by a closing of the market during the contract period—inevitably results in the disadvantage that the potentials of active competition on the lucrative lines cannot be exhausted (cf. Knieps, 2013, pp. 167f.).

Active competition between different suppliers of public transport services can lead to a coordination problem in the provision of network services. In public transport there is the problem of time table coordination. The decisions as to how often, at what time and at what quality transport services are to be supplied are the responsibility of the commissioning authority and are the subject of the auctioning procedure.

An early example is the auction for supplying transport services on the bus lines in the City of London. The first bus lines in the greater London area were auctioned as early as 1984. The design of the bus network, including all associated components of the transport services, such as frequency, types of buses used, route planning and transfer points, were determined by London Regional Transport as the commissioning authority. Only the actual provision of bus services was tendered, with the objective of cost minimisation. Both individual line sections and network areas (bundles of lines) were permitted as auction objects. Licenses were assigned for a longer period of time (as a rule for 5 years) in the form of gross cost contracts. The winner of the auction was the supplier offering transport services on a given line for the lowest compensation. The revenues from bus ticket sales went to London Regional Transport as the authority in charge of the tendering. This was one of the first implementations of a combinatorial auction which has considerable practical relevance still (cf. Cantillon & Pesendorfer, 2006, pp. 574, 589).

In the meantime competitive tendering for local bus services has been increasingly implemented in different European countries. In particular in the Scandinavian countries competitive tendering is well established in this sector, and has resulted in at times significant savings in public subsidies. Local authorities are still responsible for the central planning of lines and networks and also determine the politically desired quality standards (cf. Bekken et al., 2006).

5.3.2 Auctions of Frequencies

Natural resources, on the basis of which network infrastructures can be built, may be understood as a pool of all potential proprietary rights over land, air, outer space or water. The precondition for well-defined and thus tradable property rights is a sovereign act of structuring the potential rights (transparent definition and parcelling). Examples for this are rights over natural resources, such as frequencies and rights of way. A distinction has to be made between these property rights for natural resources and classification systems such as the metrical system or the postal code system. Classification systems have the characteristics of a public good because there is no rivalry in consumption and excluding third parties makes no economic sense. For frequencies and rights of way the situation is different, because their usage as input for building networks is associated with direct rivalry and positive opportunity costs.

In the past these natural resources were mostly allocated according to administrative rules. Only in the last decades have market-based allocation procedures been increasingly implemented. The procedure for auctioning frequencies became particularly popular. Although there were hearings on this issue in the US congress as early as 1958—and Coase explicitly endorsed auctions in his 1959 paper "The

Federal Communications Commission"—it took until 1993 for the US congress to approve auctioning of the frequency spectrum (cf. McMillan, 1994, p. 147).

In Great Britain the auctioning of third generation mobile communications licenses (UMTS) took place in April 2000, with other European countries following suit. While there was wide agreement on auctions being a more efficient allocation mechanism than so-called "beauty contests" (that is, direct contracting to "suitable" candidates), the focus was on the quest for the best auction design and the suitable auction object in the form of frequency packages (cf. e.g. Abbink et al., 2005; Klemperer, 2002, pp. 184ff.; Seifert & Ehrhart, 2005). The revenue from the auction for the state, the danger of collusion among the bidders, as well as the crowding out of market entrants, and the risk of winner's curse were of central importance. In different countries different auction forms were developed, with both the auction objects and the auction rules varying considerably.

In Great Britain five licenses with a predetermined bandwidth were auctioned, with every bidder being allowed to purchase one license at maximum. The bandwidth of licenses varied, so that the different licenses had different values. The most valuable license was reserved for a market entrant (cf. Seifert & Ehrhart, 2005, pp. 230ff.). Although the government first planned an English-Dutch auction, with a final round in which five bidders would each submit a final sealed bid, in the end it decided on a pure ascending auction (cf. Klemperer, 2002, pp. 184f.).

In Germany in contrast the number of licenses was not ex ante given. The auction objects consisted of multiple objects (frequency blocks), so that the total number of licenses and the bandwidths emerged endogenously in the course of an ascending auction (cf. Seifert & Ehrhart, 2005, pp. 232f.).

5.4 Questions

5-1: Vickrey Auctions
Explain the basic principle of the Vickrey auction and give reasons why this type of auction leads to honest bids. What role does the assumption of risk neutrality play?

5-2: Auctions and Price Differentiation
Under which conditions do auctions lead to price differentiation?

5-3: Auctions in Network Industries
Explain the role of invitations to tender in the provision of universal services.

5-4: Competitive Invitations to Tender
Explain the role of competitive invitations to tender based on the example of London bus transport.

References

Abbink, K., Irlenbusch, B., Pezanis-Christou, P., Rockenbach, B., Sadrieh, A., & Selten, R. (2005). An experimental test of design alternatives for the British 3G/UMTS auctions. *European Economic Review, 49*, 505–530.

Bekken, J.-T., Longva, F., Fearley, N., & Osland, O. (2006). Norwegian experience with tendered bus services. *European Transport/Trasporti Europei, 33*, 29–40.

Cantillon, E., & Pesendorfer, M. (2006). Auctioning bus routes: The London experience. In P. Cramton, Y. Shoham, & R. Steinberg (Eds.), *Combinatorial auctioning* (pp. 573–591). Cambridge, MA: MIT Press.

Coase, R. H. (1959). The federal communications commission. *Journal of Law and Economics, 2*, 1–40.

Cramton, P., Shoham, Y., & Steinberg, R. (2006). Introduction to combinatorial auctions. In P. Cramton, Y. Shoham, & R. Steinberg (Eds.), *Combinatorial auctions* (pp. 1–14). Cambridge, MA: MIT Press.

Demsetz, H. (1968). Why regulate utilities? *Journal of Law and Economics, 11*, 55–65.

Demsetz, H. (1969). Information and efficiency: Another viewpoint. *Journal of Law and Economics, 13*, 1–22.

Klemperer, P. (2002). What really matters in auction design. *Journal of Economic Perspective, 16*(1), 169–189.

Klemperer, P. (2004). *Auctions: Theory and practice*. Princeton; Oxford: Princeton University Press.

Knieps, G. (2006). Delimiting regulatory needs. In OECD/ECMT Round Table 129, Transport services: The limits of (De)regulation, Paris, pp. 7–31.

Knieps, G. (2013). Competition and the railroads: A European perspective. *Journal of Competition Law and Economics, 9*(1), 153–169.

Lipczynski, J., Wilson, J., & Goddard, J. (2005). *Industrial organization* (2nd ed.). Harlow: Pearson.

Lucking-Reiley, D. (2000). Vickrey auctions in practice: Nineteenth-century philately to twenty-first-century e-commerce. *Journal of Economic Perspective, 14*(3), 183–192.

McAfee, R. P., & McMillan, J. (1987). Auctions and bidding. *Journal of Economic Literature, 25*(2), 699–738.

McMillan, J. (1994). Selling spectrum rights. *Journal of Economic Perspectives, 8*, 145–162.

Milgrom, P. R. (1987). Auction theory. In T. F. Bewley (Hrsg.), *Advances in economic theory*. Cambridge: Cambridge University Press.

Nash, J. F. (1951). Non-cooperative games. *Annuals of Mathematics, 54*, 286–295.

Robinson, M. S. (1995). Collusion and the choice of auction. *Rand Journal of Economics, 16*, 141–145.

Rothkopf, M. H., Teisberg, T. J., & Kahn, E. P. (1990). Why are Vickrey auctions rare? *Journal of Political Economy, 98*(1), 94–109.

Seifert, S., & Ehrhart, K.-M. (2005). Design of the 3G spectrum auctions in the UK and Germany: An experimental investigation. *German Economic Review, 6*(2), 229–248.

Ungern-Sternberg, T. von (1991). Swiss auctions. *Economica, 58*, 341–357.

Varian, H. R. (2010). *Intermediate microeconomics* (8th ed.). New York: Norton.

Vickrey, W. (1961). Counterspeculation, auctions, and competitive sealed tenders. *Journal of Finance, 16*, 8–37.

Wolfstetter, E. (1996). Auctions – An introduction. *Journal of Economic Surveys, 10*, 367–420.

Compatibility Standards in Networks

6.1 Basic Elements

Compatibility standards have for a long time been of great importance in many areas of the economy. Examples are measures for weights and lengths, languages, the width of rail tracks, voltage, or transmission and switching protocols in telecommunications. Traditionally, setting, enforcing and changing suitable compatibility standards has primarily been the responsibility of engineers and lawyers. Only a few decades ago have economists started to address this problem.[1] In the meantime the economics of compatibility standards is a well-established field in modern industrial economics and network economics.[2]

6.1.1 Direct and Indirect Network Externalities

An important role in the economic analysis of compatibility standards is played by consumption externalities. The higher the number of individuals consuming the same good, the higher an individual's utility in consuming the same good.[3] Positive consumption externalities are usually also termed network externalities. All individuals who experience positive consumption externalities when consuming an identical or compatible product are considered to belong to the same "network" (cf. Katz & Shapiro, 1985, p. 424). A simple example of this is a telecommunications network. The larger the number of people connected to such a network,

---

[1] In the context of a case study regarding the liberalisation of telecommunication services the lack of an economic theory of compatibility standards has been pointed out in Knieps et al. (1982, p. 213).

[2] Early studies on the economics of compatibility standards are Farrell and Saloner (1985), Katz and Shapiro (1985), Kindleberger (1983).

[3] Contrary to the traditional assumption of welfare economics that one individual's utility of consumption is not dependent on the other individuals' consumption.

the higher a consumer's utility in being connected to the same network, because the possibilities of communication increase. The earliest studies examining network externalities emerged in the context of telecommunications networks. Here, the focus is on access to a telephone network rather than the problems of compatibility standards between different networks.[4] In contrast, Kindleberger (1983) does not use the term network externality in his historical analysis of the importance of compatibility standards; he only refers to the reduction of transaction costs and the advantages for production brought about by standardisation of, e.g. measures and weights or standardised widths for railway tracks. Modern industrial organisation literature differentiates between the direct physical effects on the quality of a product (e.g. direct network externalities in communications networks) and indirect network effects that also cause consumption externalities.[5]

If two communications networks are compatible with each other because of a common standard, the sum of the users of these two networks determines the extent of the direct physical network externalities. But even when consumers' utility depends exclusively on the technical characteristics of a product (e.g. a PC) and its complementary products (e.g. PC software), consumption externalities are still relevant. The quantity and variety of available software offered in the market is dependent on the number of existing compatible or identical hardware units. Increasing economies of scale in the production of complementary products can thus have the same effect as the existence of direct physical network externalities. An increase in the number of consumers of (software-) compatible hardware leads to economies of scale in the production of the relevant software supply and increases each individual's consumer surplus (cf. Chou & Shy, 1990). It is not only in the context of hardware/software (computer software, video cassettes) that these indirect network effects become apparent, but also in the density of complementary service networks for long-lasting economic goods (e.g. car repair shops). If there is a choice between cars with different fuel technologies (e.g. petrol versus natural gas), the consumer's decision for a car with a specific fuel technology (e.g. natural gas) also depends on the density of the network of filling stations where natural gas is available (cf. Conrad, 2006).

6.1.2 Standards as Public Goods, Private Goods, and Club Goods

Standards can have the character of public goods, private goods, or collective goods,[6] depending on the interplay of network effect and technology effect.

– The technology effect describes the utility to be derived from a specific technology.

---

[4] Cf. e.g. Artle and Averous (1973), Rohlfs (1974), Oren and Smith (1981).
[5] Cf. e.g. Katz and Shapiro (1985), as well as Farrell and Saloner (1985, 1986a).
[6] Cf. Besen and Saloner (1989), Berg (1989), Kindleberger (1983).

– The network effect describes the utility resulting from the number of economic agents using the same standard.

Consider the standards for measures, whether measures of length, weight, temperature, time or value. Economic agents are relatively indifferent as to which standard is realised and have no preference for a specific concretisation of the standard. Their incentives to get involved in the process of standard-setting are therefore minimal, even though all would profit from the establishment of a unified standard and the resultant network externalities. Every economic agent will attempt to avoid the cost of standard-setting (this is the so-called free rider phenomenon). In these cases compatibility cannot be achieved in a decentralised way through the market. Instead, the standard takes on the character of a public good which must be provided by the state, if it is to be provided at all.

In contrast to standards, which are purely public goods, there are technical specifications that exhibit primarily the characteristics of private goods. There is a strong preference for using different technologies simultaneously. This case occurs when the technology effect is so large that it outweighs the network effect, even if there are only few users belonging to the network. If the network effect is zero in the extreme case, the situation is that of the common microeconomic case without positive network externalities. Here, incompatibility between end products is the state preferred by all.[7] For example, standards for motor vehicles, household appliances or industrial plants produced by different manufacturers have the character of a private good, because they serve the objective of guaranteeing a certain standard of quality.

The third category are standards that can be viewed as collective goods. The starting point is a situation in which the advantages of compatibility are seen as sufficiently large so that an active participation—e.g. of the businesses in one industry—in the standardisation process can be expected (so that there is no free-rider problem). For example, nowadays different industries have their own specific telecommunications standards.[8] As the preferences for a specific technology may still vary considerably between different economic agents in one economy, there is a possibility of network islands emerging for different types of demand (cf. e.g. Farrell & Saloner, 1986b).[9] In such a case network externalities are consciously abandoned in favour of the variety of different technologies. In

---

[7] This does, however, not preclude the possibility that producers may have incentives to only supply compatible components to the market.

[8] For example, SWIFT (Society for Worldwide Interbank Financial Telecommunications) is used by banks. SWIFT is a high-quality service network based on a secure communication standard for the electronic exchange of standardised SWIFT messages (cf. Knieps, 2006a, p. 55). EDIFACT (Electronic Data Interexchange for Administration, Commerce and Transport) is used worldwide by the consumer goods industry. It is a standard for network services applied for structured information of business correspondence (cf. Blankart & Knieps, 1995, p. 293).

[9] The extreme case of one standard being accepted by all economic agents is of course not excluded.

Table 6.1 Standards as public goods, private goods and club goods

Network effect → Technology effect ↓	Large	Small
Large	Standards as club goods (conflict of interests)	Standards as private goods
Small	Standards as club goods (coordination problems)	Standards as public goods

addition to the case where all technologies in an industry are incompatible and the case where all technologies in an industry are compatible, there is the possibility of firms building standard coalitions, so that the technologies of all the members of one coalition are compatible (cf. Economides & Skrzypacz, 2003).

The extreme case of all economic agents being indifferent as to which standard is adopted—as is, for instance, the case for standards for measures (length, weight, temperature)—so that the standard takes on the character of a public good, will not be considered in the following. Nor will we examine the situation of the common microeconomic case without positive network externalities where products by different firms are completely incompatible. Instead, in the following, compatibility standards are regarded as club goods, so that the users of an industry standard create a positive network externality for all users of a technology compatible with this standard. Nevertheless conflicts of interests are unavoidable, if the technology effect is large. If the technology effect is small, the focus is on the coordination problem. Table 6.1 provides an overview of the differentiation of standards for the different types of goods.

6.1.3 Network Externalities Between Network Variety and the Search for New Technologies

Network externalities can be formalised as follows:

$$u_i = u_i(S, T) \tag{6.1}$$

that is, the utility u_i for an individual i to be connected to a network is dependent on the network's technology T and the total number of participants S connected to this network.[10] If there are positive network externalities, then:

---

[10] For reasons of simplification, in the following the set of participants is assumed to be equal to the number of participants.

$$u_i(S,T) < u_i(S',T) \quad \text{for } S < S'. \tag{6.2}$$

With this definition, congestion effects in networks (i.e. negative network externalities) are excluded (cf. Chap. 3).

Because of network externalities it is advantageous to choose the most compatible network technologies available. The greater the number of individuals using the same network, the higher the utility. But the validity of this principle is limited by the differences in individual preferences. For the totality of individuals N it might under certain circumstances be more advantageous to split up into several, non-compatible network islands than to form one large unified network (cf. e.g. Farrell & Saloner, 1986b). Such network islands can for example develop if, after the emergence of a new technology T_2, it is for some individuals more advantageous to stick to the old technology T_1, while for others it is more profitable to switch to the new technology T_2.

$$u_i(S,T_1) > u_i(N,T_2), \quad i \in S \tag{6.3}$$

$$u_i(\widetilde{S},T_2) > u_i(N,T_1), \quad i \in \widetilde{S}, \tag{6.4}$$

where $S + \widetilde{S} = N$.

In a scenario of comparative statics technology T_1 is compared to technology T_2. However, in a dynamic world the set of possible technologies $\{T_1, T_2, \ldots\}$ is open. The number of its elements cannot be determined ex ante, because the technologies must first be discovered, before they can be implemented (cf. Blankart & Knieps, 1993b, pp. 44ff.).

6.1.4 Standards for Goods, Complementary Components and Large Technical Systems

Many studies on compatibility standards examine the standardisation of single products. In this context, compatibility means that products exist in the same "network" with regard to their technical specifications.

Even if no positive network externalities exist, producers may find themselves faced with the decision whether they should produce individual components that are compatible with the components produced by other manufacturers, or rather create incompatible systems (cf. Economides, 1989; Matutes & Regibeau, 1988). This is a situation where consumers have a preference for using a system made up of a set of compatible elements. In this context, compatibility means that a consumer is able to combine components produced by different manufacturers, resulting in a greater variety of goods; this enables the consumer to use the specific version of the system that he or she prefers (cf. Holler, Knieps, & Niskanen, 1997, pp. 386ff.).

In the large technical systems of network industries it is necessary to approach the problem in a disaggregated fashion. After the comprehensive opening of railway, air traffic, energy and telecommunications networks, fully integrated, hierarchical standardisation has been replaced by, on the one hand, standardising interfaces between individual network levels (vertical standardisation problem) and, on the other hand, standardisation within individual network levels (horizontal standardisation problem).

Network externalities are less important on the level of network services than on the level of network infrastructures. A multitude of non-compatible service networks may emerge, because customers have preferences for different network services. Although the variety of different network islands is crucial, incentives for compatibility standards between individual service network components may nevertheless exist. An illustrative example for this is the traditional Internet. Internet service providers offer their customers a multitude of different services, so that a large number of different service networks develop. Still, all these service networks are based on a common network logistics which is in turn based on the technical standard of the TCP/IP. While the IP (Internet Protocol) is responsible for directing the data to the correct recipient, the TCP (Transmission Control Protocol) is in charge of the reliability of the transmission (cf. e.g. Knieps, 2003, p. 224). In the meantime, due to the transition from narrowband to broadband Internet, the traditional best-effort TCP/IP-based Internet assigning all data packets equal priority is being challenged. Heterogeneous traffic qualities within different service networks become increasingly important, creating the need for quality of service-based interconnection agreements between different Internet service providers (cf. Knieps, 2011, pp. 25, 36).

On the level of network infrastructure, network externalities and compatibility between different networks are of considerable importance. Cross-border railway traffic, for example, requires that the track gauges match (cf. e.g. Blankart & Knieps, 1993a, pp. 49f.). In the meantime, the goal of trans-European railroad networks has created a new momentum to support the interoperability of the railroad systems across European countries (cf. Knieps, 2013a, p. 164). In the same way, different electricity networks require the same voltage for transmission between interconnected networks. The process of integrating renewable energy sources into the European market for electricity highlights this as the focus is shifted from national electricity networks to cross-border compatible infrastructures (cf. Knieps, 2013b, pp. 147ff.).

6.2 The Coordination Problem

6.2.1 Spontaneous Switching to a Superior Technology

The coordination problem can be illustrated by means of a comparatively simple case in which only the transition from an old technology (e.g. Telex) to a new, superior technology (e.g. Telefax) with advantages for all users is examined. The possibility of a transition to a third technology will be excluded at this point. It is assumed that the technology effect is weaker than the network effect and only guarantees that every user will profit from the new technology if all other users

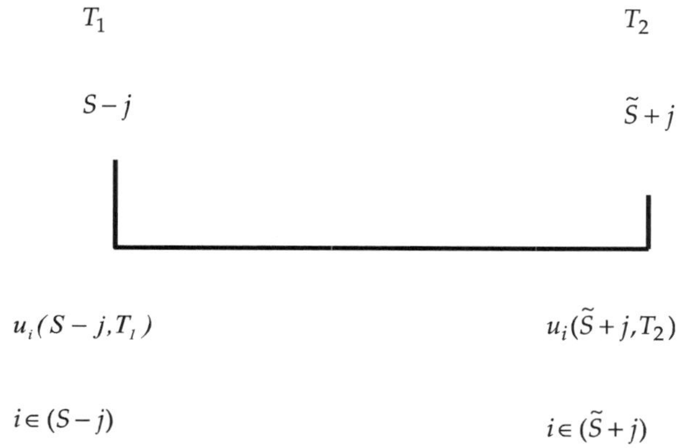

Fig. 6.1 Utility effects of the transition to a new network

T_1 T_2

$S-j$ $\tilde{S}+j$

$u_i(S-j, T_1)$ $u_i(\tilde{S}+j, T_2)$

$i \in (S-j)$ $i \in (\tilde{S}+j)$

decide to switch, too. Only under particular conditions can the market initiate this switch spontaneously (cf. Farrell & Saloner, 1985).

Assume that there is a choice between two networks T_1 and T_2. Let T_1 be the old, traditional network and T_2 a new network. Let N denote the number of potential network participants that can choose between T_1 and T_2. A number of participants S are employing technology T_1 and a number $\tilde{S}$ are employing technology T_2.

It is assumed that each of the N individuals is either part of the old or of the new network, that is $S + \tilde{S} = N$. If an individual j switches form network T_1 to the new network T_2, this creates a decrease in externality (loss of utility) for the group using T_1 and an increase in externality (utility gain) for the group using T_2 (cf. Fig. 6.1).

How do these externalities influence the calculation of individual j, faced with the decision if he or she should switch from the old to the new network? On the one hand, the economic agent will assess the technology effect, i.e. the utility of the technology inherent in network T_2 versus the utility of the technology inherent in network T_1. In addition, he or she will take into account the network effect, i.e. the utility resulting from the number of individuals $\tilde{S} + j$, which will be members of the new network after the switch. The fact that the number of individuals in network T_1 will be reduced by one does not enter this calculation. A decrease in externality for the users of the old network is created. An individual j will switch to the new network, if:

$$u_j(S, T_1) < u_j\left(\tilde{S} + j, T_2\right) \tag{6.5}$$

Regarding the decision to switch, the following cases can be differentiated:

On the one hand the technology effect can be so strong that it outweighs the network effect, even if only very few users switch to the new network, i.e. if $\tilde{S}$ is relatively small. If the network effect is zero, this represents the common microeconomic case without positive network externalities. If, however, on the other hand the technology effect is weaker as compared to the network effect, so that an individual user can only be sure to benefit from a switch if all other users switch as well, i.e. if:

$$u_i(N, T_2) > u_i(N, T_1), i \in N \tag{6.6}$$

the question arises whether the switch to the new network will occur spontaneously.

Under the following conditions a spontaneous switch is assured:

- no irreversible costs;
- perfect information regarding the individuals' preferences, i.e. each individual knows that switching is profitable for all if all individuals switch, and that this fact is known to all;
- no strategic behaviour, i.e. the sequence of switching is irrelevant for each individual's utility calculation;
- there are only two technologies.

This can be demonstrated by the following inductive argument (cf. Farrell & Saloner, 1985, Sect. 2):

If out of N individuals all but the Nth have adopted the new (superior) technology, the Nth individual also has an incentive to do so, because:

$$u_N[(N-1) + 1, T_2] > u_N(N, T_1) > u_N(1, T_1) \tag{6.7}$$

From this it follows that the $(N-1)$th user can be sure that the Nth will follow him, provided that all precursors have already joined the new network. Therefore he, too, will switch to the new network, and so on. The same argument holds for all other users, down to the very first user. Given the assumptions listed above, each one will find himself in the position of the Nth user and switch to the new network. Thus it will not only be some of the individuals that join the new network, but the entire group. Under these conditions the switch to the new technology will occur spontaneously and without coordination. A generalisation of these results to include cases of more than two technologies is possible, as long as there is one technology that is globally preferred over all others.

6.2.2 The Phenomenon of Critical Mass

If the conditions for a spontaneous switch from an existing to a new, superior technology are not fulfilled, there will be the problem of critical mass, i.e. of the minimum number of participants needed to make the new network self-sustaining.[11] This problem is relevant for the introduction of a new technology as well as

---

[11] In addition there is the problem of the discrepancy between the private and the social incentives as to what constitutes optimal network size, as well as the problem of network fragmentation (cf. Sect. 6.3.2).

for the switch from an existing technology to a new one. The concept of critical mass was originally developed in the context of analysing the structure of telecommunications networks. In this context, critical mass is defined as the smallest number of participants in a network that, with a uniform minimal connection charge, make a network cost-covering (cf. e.g. Oren & Smith, 1981, p. 472). A formal analysis of this problem is given in Rohlfs (1974, p. 29). As soon as the number of participants in a network exceeds critical mass, spontaneous network development can occur. The concept of critical mass has proved to be highly useful in the more general context of moving towards a new technology.

6.2.3 Path Dependency

If there is imperfect information on the utility of the network externalities for the other economic agents, nobody can be sure that the others will follow him or her. This uncertainty can lead to a situation where all individuals wait for a large enough number of the others to start switching. This is the so-called penguin effect: penguins hovering on the edge of the ice attempt to let the others jump first into the water; although they are all hungry, each fears that there might be a carnivore fish in the water (cf. Farrell & Saloner, 1987, pp. 13f.). The consequence may be a persistence of the status quo, even though all would prefer a switch to the new technology. Considering various examples, economic historians have pointed out the relevance of such scenarios and referred to this phenomenon as path dependency (cf. e.g. Arthur, 1984; David, 1985).[12]

An interesting example is the long persistence of the programming languages FORTRAN and COBOL: Despite the development of new programming languages (e.g. C++ or Java), which are for many purposes more user-friendly, offer greater variety of application, and are in general associated with considerably lower programming cost, FORTRAN and COBOL are still in use in many areas.[13]

The best-known example is the so-called QWERTY keyboard for typewriters and PCs. The term QWERTY refers to the sequence of letters in the top row of this type of keyboard. According to David (1985) this standard asserted itself due to historical reasons which from the point of view of efficiency are no longer relevant today: In the early days of the typewriter when the technology was not yet fully developed, the individual keys had to be arranged in such a way that the type bars did not become jammed during fast typing, i.e. during the process of typing keys

---

[12] In the natural sciences path dependency has long been known as hysteresis. It refers to the dependency of the physical state of an object on the preceding states. For example, when a piece of iron, magnetised to the point of satiation, is degaussed per slow reduction of the field intensity, a residual magnetisation, the so called remanence, remains.

[13] It should be noted, however, that these traditional computer programming languages also have undergone significant changes over time, improving their performance characteristics. For the final draft of the Fortran 2008 standard, see Reid (2008). For the COBOL 2002 standard see Oliveira (2006).

worked with the right hand should alternate with keys worked with the left hand as often as possible. It is said that the QWERTY sequence of the keyboard conformed best to this objective. Over time, typewriter technology improved and the QWERTY keyboard did no longer constitute a technical necessity. As an ergonomically adapted improvement, the DSK keyboard was developed. Experiments are said to have shown that with this keyboard typing at 20–40 % increased speed is possible (cf. David, 1985, p. 332). But this innovation never took hold, because the QWERTY keyboard was already established. Users (allegedly) did not find it worthwhile to learn DSK, because DSK typewriters could not be bought, while typewriter manufacturers considered the production of DSK typewriters to be unprofitable, because nobody could type on these keyboards. Thus, users and typewriter manufacturers stuck with the old QWERTY system. The QWERTY example started an important scientific debate on the dangers of the persistence of the status quo due to the problems of coordinating user externalities.[14]

6.3 Conflicts of Interest

6.3.1 Producers

In the literature examining the incentives for standard-setting, oligopoly models with a predetermined number of producers play an important role. If, due to positive network externalities, consumers benefit from compatible products by different producers, the question arises why producers should not also have an interest in their products being compatible. Under perfect competition the setting of compatibility standards would only be a coordination problem. Producers could not gain strategic advantages from incompatibility.

In the more realistic case of a smaller number of producers being active in the market the picture changes. If a firm decides to make its product compatible with the products of other firms, this is now a parameter of behaviour which can be used strategically like price setting behaviour, product choice or advertising. When examining the firms' incentives to create compatible versus incompatible products, the traditional industrial economics approach of assuming a causal nexus between structure, conduct and performance does not yield any useful results. In particular, it is not possible to derive a general statement as to the structural market conditions under which an optimal degree of standardisation will be achieved. Instead, the analysis of a firm's incentives for compatibility by means of game theoretical approaches proves to be more promising. In the meantime, a multitude of models

---

[14] However, the inefficiency of the QWERTY standard has meanwhile come under debate; it is said that at the time it was at least among the acceptable alternatives (cf. Liebowitz & Margolis, 1990). Even if the QWERTY example has lost its empirical relevance, it serves as a particularly clear illustration of the problems of coordination inherent in introducing new standards.

for analysing strategic pricing and non-pricing behaviours have been developed. In this context, various models of standardisation have also been examined.[15]

An important cognitive value of these game theoretical models is their ability, given specific model assumptions, to consistently derive the conflict potentials between different producers. Thus it has been shown for the case of an asymmetric duopoly equilibrium that a dominant firm has no interest in the compatibility of its technology, if compatibility reduces the value of its product in comparison to that of the smaller supplier and lessens its profit (cf. Katz & Shapiro, 1985). If positive network externalities exist, this is quite plausible, because the customers of the dominant supplier already benefit from the substantial network externalities of the large group of compatible customers, whereas the customers of the small supplier will only be able to benefit from substantial network externalities after the adoption of a compatibility standard.

Other forms of strategic behaviour can also be explained by means of such game theoretical approaches. For example, by means of strategic price setting behaviour the monopolist of an old technology can under specific conditions prevent the adoption of a new technology under competition, even if the new technology is superior for all users (cf. Besen & Saloner, 1989; Farrell & Saloner, 1986a). The effects of announcing a new technology can also be relevant in the competition between technologies (cf. e.g. Dranove & Gandal, 2003).

While for example Arthur (1983, 1984) and David (1985) emphasise the role of historic coincidence leading to the emergence of standards which subsequently persist for extended periods due to positive network externalities, Katz und Shapiro (1985, 1986) attribute the persistence of standards to market power. They attempt to explain the emergence of standards under oligopolistic competition. The results are strongly dependent on the original market shares and on expectations. It is hardly possible to derive general statements regarding the strategic incentives of firms in the standardisation process from these models.[16]

6.3.2 Consumers

In the following we will assume that there is a new and an old technology; it is not known which of the two is better and whether the new technology T_2 is superior to the old technology T_1, when it is accepted by all individuals, i.e. when the network effect is fully exhausted. In this case it is beneficial to behave strategically. It becomes important which individual switches first and which switches later, and inefficient persistence of the status quo is possible, because each individual wants the others to switch first.[17] Those wishing to switch are unsure whether a

---

[15] For this cf. Gilbert (1992) as well as Besen and Farrell (1994).

[16] For this and similar studies cf. Katz and Shapiro (1994).

[17] Cf. Farrell and Saloner (1986a). They also discuss the case of inefficient mobility, where switching occurs too fast, because nobody wants to be left behind alone with the old technology. Considering that there are frequently costs involved in switching to the new technology, this case does not seem directly relevant.

sufficiently large number of individuals will follow them to make the switch profitable; thus the problem of critical mass arises. In particular, it is not clear whether

$$u_i\left(\widetilde{S}, T_2\right) > u_i(S, T_1) \quad \text{where} \quad \widetilde{S} + S = N \tag{6.8}$$

does apply. The question is whether there are enough individuals for whom this holds true. Only then will critical mass be reached and the new technology be adopted. Otherwise the new technology will not be adopted, even if it proves superior.

The problem of critical mass also occurs if, in addition to technology T_1, two possible new technologies T_2 and T_3 are available, both superior to T_1 if all users switch to them. If it is true for these two new technologies T_2 and T_3 that:

$$u_i(S, T_2) > u_i(N, T_3) \tag{6.9}$$

and

$$u_i\left(\widetilde{S}, T_3\right) > u_i(N, T_2), \quad S + \widetilde{S} = N \tag{6.10}$$

then it follows that in this case the users will profit more from splitting off into several different network islands than from realising the advantages of the network externalities of one large common network (cf. e.g. Farrell & Saloner, 1986b). But even in the absence of strategic behaviour it is not certain that fragmentation, for example into two different networks, will occur. It is also possible that none of the new technologies will reach critical mass and users will continue working with the inferior status quo technology.

6.4 Standard-Setting Institutions

6.4.1 Government Intervention

It needs to be examined whether and to what extent the existence of positive network externalities—and the concomitant problem of critical mass—is sufficient to derive the necessity of government intervention. At the same time, the preferences of economic agents for different technologies have to be taken into account. After all, government intervention must not obstruct the search for new technologies.

In order to reach critical mass it is conceivable to subsidise the costs of all new technologies and thus promote their financial viability. As a consequence, the

6.4 Standard-Setting Institutions

required critical mass would be smaller. In the extreme case, all network islands might become financially viable. Alternatively, subsidies may be distributed selectively, so that at least some technologies will reach critical mass. And finally, it would be possible to develop a specific technology policy and only permit a small number of selected technologies on the market, so that a sufficiently large number of individuals using them is guaranteed.

Subsidies and an active technology policy in fact lead to insurmountable knowledge problems. Weighing the economic cost against the utility of a specific course of action would completely overcharge the capacity of a standardisation authority. This is because network externalities lead to pronounced nonlinearities, typically resulting in a multitude of non-comparable Nash equilibria with different numbers of network islands.[18] However, in a dynamic world the problem of achieving the optimal number of network islands with different, non-compatible technologies is unsolvable on principle. As the number of technologies is unknown, it is impossible to calculate the optimal number of network islands ex ante. Evolutionary economics rightly emphasises that the persistence of the status quo technology, even if it is inferior, may well be caused by the fact that nobody can predict with certainty which qualities alternative technological solutions will exhibit in the future (cf. Witt, 1991). The development of an institutional approach based on the optimisation of network externalities, variety and searching is therefore impossible.

If standards are established by government legislative authority, this does not mean that the legislators will handle the standardisation process themselves; instead this task is delegated to the relevant institutions. In doing so, a bureaucracy will, within the margins of its freedom of action, maximise its own utility (budget or output), which leads to the risk of excessive standardisation (cf. Blankart & Knieps, 1993a, p. 46).

Under this perspective, many profound, highly specialised standards established by bureaucratic institutions must be viewed with a certain amount of scepticism. An illustrative example of government interventionism was the debate on the implementation of the EDIFACT (Electronic Data Interchange for Administration, Commerce and Transport) standard by government decree. For structured information (such as bills, order forms, delivery notes etc.) to be processed directly by the recipient's computer, each data field must be assigned a clearly defined significance. Government promotion of this highly specialised standard ensures a large user base, but at the same time strongly impedes the emergence of network islands and competing standards, that is, variety and innovation.

A technology policy that only permits certain selected technical systems could only be efficient if the superiority of the chosen technologies was certain and the search for alternative technologies did not present any problems. In this world of perfect information, the objectives of building networks rapidly and overcoming

---

[18] Under very specific assumptions (e.g. two consumer groups, two goods) specific results regarding the trade-off between network externalities and variety can be derived (cf. Farrell & Saloner, 1986b).

the problem of critical mass could be easily achieved by interventionist means. The network provider would conduct a subscription and only build the network after a sufficiently large number of users had signed on. However, it is not known what the best network technology might be, and new solutions emerge continuously; therefore interventionism is a questionable approach, because it commits users to a technology that might prove to be inferior and possibly not even be able to reach critical mass.

6.4.2 Market Solutions

Even if network externalities exist, (uncoordinated) market solutions play an important role in the development and implementation of standards. These are de facto standards, without any involvement by standard-setting institutions. Examples of this phenomenon are standards set by one firm which spontaneously assert themselves on the market to become industry standards.

6.4.2.1 Network Evolution Under Monopolistic Competition

From the theory of monopolistic competition it is well known that the trade-off between economies of scale and product variety is best left to market processes, while government intervention would lack the required information to implement social welfare maximising solutions.[19] This holds true, even if the socially optimal product variety does not emerge, as corrective measures on the part of the government (taxation and subsidisation of individual products) may lead to welfare reductions. In particular, the conflict between network externalities, variety and the search for new technologies cannot be solved by interventionist means (cf. Blankart & Knieps, 1993b). The historical development of competing, incompatible technologies is known from various economic fields. Examples are e.g. different video systems produced by different firms, different operating systems and word processing programs for PCs or different electricity transmission systems (DC versus AC). Direct and indirect network externalities can result in one single technology prevailing in the end, thus establishing a de facto industry standard (cf. e.g. Shapiro & Varian, 1999b).

6.4.2.2 Gateways

The weighing of network externalities versus network variety is made easier by the development of gateways that enable compatibility between different network services. An example of gateways for network infrastructures is the installation of interconnectors between electricity networks.

Gateway technologies serve the function of making different technologies at least partly compatible. They can be found in many industries, e.g. as adaptors for different power sockets and currents. So to what extent does market entry via

---

[19] An overview of this theory can be found in Carlton and Perloff (2005, Chap. 7).

gateway technologies reduce or even eliminate the problem of incompatibility between different technologies? As converters are associated with costs and the full exploitation of all network externalities is often not possible, the advantages of complete compatibility without any loss in variety cannot be achieved via converters either (cf. Farrell & Saloner, 1992, p. 32). However, the endogenous emergence of gateways on the market can reduce the problem of incompatible technologies significantly (cf. Blankart & Knieps, 1993b, p. 47). A converter between two different technologies retains the advantages of technology variety, while at the same time at least partly obtaining the advantages of network externalities. There are one-way and two-way converters. Two-way converters enable a (partial) compatibility in both directions, whereas one-way converters make the network externalities of both technologies available to only one user group (cf. Farrell & Saloner, 1992, p. 12).

In the context of the evolutionary development of competing technologies the market-compatible development of a gateway technology can lead to an outcome where one single technology prevails. An illustrative example is the competition between DC and AC networks (cf. David & Bunn, 1988; Shapiro & Varian, 1999a, pp. 210ff.). The development of the rotary converter enabled the combination of long distance (high voltage) AC networks and regional DC distribution networks, so that it became possible to profit from the network externalities of DC technology. Thus the rotary converter contributed significantly to the ultimate success of AC electricity transmission.[20]

If it is already known that a converter exists before the decision for one technology is made, the two-way converter can render the switch from an existing to a new technology easier but also more difficult. On the one hand, the incentives for strategic waiting (penguin effect) are reduced; those that have switched to the new technology can still (at least to some extent) benefit from the network effects of the old technology, so that the small user base of the new technology at the beginning does not matter. This increases the new technology's chances of reaching critical mass. However, on the other hand there is the possibility that the existence of converters increases the incentives of persisting in the status quo, as there is no danger of the existing technology becoming worthless due to a lack of network externalities (cf. Choi, 1996).

6.4.3 Committee Solutions

6.4.3.1 The Pure Coordination Problem

While in case of market solutions, economic agents get involved in the process of standardisation in an uncoordinated manner, committee solutions are characterised by coordination and communication within the standardisation process. Standard

---

[20] A detailed analysis of the competition between these two transmission systems can be found in David and Bunn (1988); cf. also Farrell and Saloner (1992, p. 15).

committees are better able to solve the problem of coordination and, consequently, the problems of critical mass and path dependency. If there is merely an information problem regarding a technology that is superior for all, the penguin effect can be overcome by means of coordination and communication. Committees thus enable the introduction of a socially optimal standard.

Even if individuals come to know their preferences better over time, many standardisation institutions regard their task as a typical coordination activity with the objective of creating a consensus between the parties involved by means of a communication exchange, so that the persistence of the status quo can be overcome. Examples of important non-governmental organisations that publish standards in the form of recommendations are the Deutsches Institut für Normung (DIN), the American National Standards Institute (ANSI) and the International Standards Organisation (ISO) (cf. Besen & Saloner, 1989).

6.4.3.2 Conflicts of Interests

Committees, too, face difficult decision processes, if heterogeneous preferences for network variety create incentive problems because of conflicts of interest, or if standard-setting involves considerable costs (cf. Blankart & Knieps, 1993a, pp. 45f.). Decision-making costs can occur when compromises have to be made among the members of a committee regarding the nature and the extent of standardisation. Free riding behaviour by delaying the introduction of a standard is most likely to occur among those members of a committee that expect few benefits from the standard.

An explicit model comparison between market solutions and committee solutions for the special case of two user groups and two different technologies can be found in Farrell and Saloner (1988). They assume that each user group prefers a different technology, but still has a strong preference for compatibility. Thus both user groups would rather have an agreement and forego their preferred technology than deal with a state of incompatibility.

Let T_1 and T_2 be the alternative technologies, and S and $\tilde{S}$ be the two user groups $S + \tilde{S} = N$, then:

$$u_i(S, T_1) < u_i(N, T_2) \quad i \in S \tag{6.11}$$

$$u_i(\tilde{S}, T_2) < u_i(N, T_1) \quad i \in \tilde{S} \tag{6.12}$$

The possibilities of compromising, or agreeing on a third technology, are excluded in this model. In spite of the strong incentive to agree on one of the two standards, the standardisation process will be drawn-out and complex. While market and committee solutions work equally well as long as time is irrelevant, committee solutions are more likely to be successful under pressure of time.[21] An important

---

[21] However, this result may be reversed if the relevant periods of time are considerably shorter for decisions made within the market process rather than by a committee.

function of committees is the search for compromise solutions during the standardisation process. Which solution prevails in a committee is strongly dependent on the members the committee is composed of, on the possibility of new members being admitted, and on the voting modalities. An interesting example can be found in the German telecommunications sector: When there was still a state monopoly in Germany, a small number of equipment manufacturers formed a closed committee (the so-called "Fachausschuss des Zentralverbandes der elektrotechnischen Industrie") and, working together with the state monopolist Deutsche Bundespost, set all compatibility standards. Market entrants had no chance to take part in this standardisation process and the introduction of technological innovations was severely impeded (cf. Knieps, Müller, & von Weizsäcker, 1982, pp. 207f.). Since the comprehensive opening of the markets for telecommunications, the standardisation process has been opened for all suppliers (service providers, network infrastructure providers, manufacturers of equipment) and for user organisations (cf. Knieps, 1995, pp. 288ff.).

The introduction of open committees, however, does by no means lead to the disappearance of conflicts of interests in the standardisation process. Based on the objective of the setting of a formal standard within a voluntary organisation for standard-setting by voting, conflicts of interest, due to heterogeneous preferences with regard to the projected standard, can influence both the length of the decision process, as well as the form it takes. Alternative forms of organisation (e.g. Open Source Software) beyond committee solutions and without explicit voting procedures may become increasingly important (cf. e.g. Simcoe, 2012; Knieps, 2013c).

6.5 Standardisation of Technical Regulatory Functions

The starting point of technical regulatory functions (e.g. postal code systems, telephone number administration, land registers) are problems of organisation, coordination and allocation that precede the provision of network services and the building of network infrastructures. Technical regulatory functions are indispensable for building and operating networks and therefore must be supplied on a non-discriminatory basis.

Technical regulatory functions can be relevant at every network level. For instance, it is necessary for the provision of transport services that the vehicles in use conform to the relevant technical safety standards and that they are periodically monitored (TÜV etc.). The providers of telecommunications services must have the technical ability to bill their customers; for this they need access to the necessary data of the participants (name, address).

Technical regulatory functions differ from network services insofar as the organisation and coordination problems to be solved concern the entire relevant market and not just one supplier. Thus they do not deal with the logistics problems of a single supplier or group of suppliers, but with the coordination of the totality of services offered on the whole relevant market (cf. Knieps, 2006a, pp. 56f.).

Technical regulatory functions are particularly relevant for infrastructure management. For instance, railway traffic control or air traffic control require a clear-cut geographic delimitation of monitoring authority and an unequivocal definition of exactly who is exercising this authority during a given time period; this constitutes a technical regulatory function. In contrast, the actual control functions in the context of capacity and safety management are to be allocated to the level of infrastructure management and can be periodically contracted out. In contrast, code sharing, interlining and joint frequent flyer programmes do not constitute technical regulatory functions.

Because of air traffic safety considerations and the allocation of transit flight rights it is necessary for air traffic control systems to be operated in a geographically delimited area by a (public or private) institution with executive authority. The relevant geographical areas do not have to correspond to individual countries, in particular as European air traffic, due to the relatively small size of the countries, has traditionally had a strong cross-border orientation. Nevertheless, in the past each country in Europe developed its own air traffic control system on the basis of national requirements. Consequently, due to the lack of compatibility standards, a multitude of different systems, specifications and procedures emerged in the European countries (cf. Knieps, 2006b, pp. 15f.).

6.6 Questions

6-1: Network Externalities
Explain the concept of network externalities, differentiating between direct and indirect network externalities.

6-2: Network Externalities and Network Variety
What is the role of network externalities in the characterisation of standards as club goods?

6-3: Critical Mass
Explain the problem of critical mass and discuss the danger, inherent in government intervention, of supporting the wrong technology.

6-4: Committee Solutions
What is the role played by committee solutions in case of heterogeneous preferences for network variety?

References

Arthur, W. B. (1983). *Competing technologies and lock-in by historical small events: the dynamics of allocation under increasing returns.* International Institute for Applied Systems Analysis

References

Paper WP-83-92, Laxenburg, Austria (Center for Economic Policy Research, Paper 43, Stanford).

Arthur, W. B. (1984). *Competing technologies and economic prediction.* Options, International Institute for Applied Systems Analysis, Laxenburg, Austria, No. 2, 10–13.

Artle, R., & Averous, C. (1973). The telephone system as a public good: Static and dynamic aspects. *Bell Journal of Economics, 4*(1), 89–100.

Berg, S. V. (1989). The production of compatibility: Technical standards as collective goods. *Kyklos, 42*, 361–383.

Besen, S. M., & Farrell, J. (1994). Choosing how to compete: Strategies and tactics in standardization. *The Journal of Economic Perspectives, 8*(2), 117–131.

Besen, S. M., & Saloner, G. (1989). The economics of telecommunications standards. In R. Crandall & K. Flamm (Eds.), *Changing the rules: Technological change, international competition and regulation in communications* (pp. 177–220). Washington, DC: The Brookings Institution Press.

Blankart, C. B., & Knieps, G. (1993a). State and standards. *Public Choice, 77*, 39–52.

Blankart, C. B., & Knieps, G. (1993b). Network evolution. In H.-J. Wagener (Ed.), *On the theory and policy of systemic change* (pp. 43–50). Heidelberg: Physica.

Blankart, C. B., & Knieps, G. (1995). Market-oriented open network provision. *Information Economics and Policy, 7*, 283–296.

Carlton, D. W., & Perloff, J. M. (2005). *Modern industrial organization* (4th ed.). Boston: Pearson; Addision Wesley.

Choi, J. P. (1996). Do converters facilitate the transition to a new incompatible technology? A dynamic analysis of converters. *International Journal of Industrial Organization, 14*, 825–835.

Chou, C., & Shy, O. (1990). Network effects without network externalities. *International Journal of Industrial Organization, 8*, 259–270.

Conrad, K. (2006). Price competition and product differentiation when goods have network effects. *German Economic Review, 7*(3), 339–361.

David, P. A. (1985). Clio and the economics of QWERTY. *The American Economic Review, 75* (2), 332–337 (Papers and proceedings of the ninety-seventh annual meeting of the American Economic Association).

David, P. A., & Bunn, J. A. (1988). The economics of gateway technologies and network evolution: Lessons from electricity supply history. *Information Economics and Policy, 3*, 165–202.

Dranove, D., & Gandal, N. (2003). The DVD-vs.-DIVX standard war: Empirical evidence of network effects and preannouncement effects. *Journal of Economics and Management Strategy, 12*(3), 363–386.

Economides, N. (1989). Desirability of compatibility in the absence of network externalities. *American Economic Review, 79*(5), 1165–1181.

Economides, N., & Skrzypacz, A. (2003). Standard coalitions formation and market structure in network industries. Working Paper no. EC-03-08, Stern School of Business, N.Y.U.

Farrell, J., & Saloner, G. (1985). Standardization, compatibility, and innovation. *Rand Journal of Economics, 16*(1), 70–83.

Farrell, J., & Saloner, G. (1986a). Installed base and compatibility: Innovation, product preannouncements and predation. *American Economic Review, 76*(5), 940–955.

Farrell, J., & Saloner, G. (1986b). Standardization and variety. *Economic Letters, 20*, 71–74.

Farrell, J., & Saloner, G. (1987). Competition, compatibility and standards: The economics of horses, penguins and lemmings. In H. Landis Gabel (Ed.), *Product standardization and competitive strategy* (pp. 1–21). North-Holland: Elsevier.

Farrell, J., & Saloner, G. (1988). Coordination through committees and markets. *Rand Journal of Economics, 19*(2), 235–252.

Farrell, J., & Saloner, G. (1992). Converters, compatibility, and the control of interfaces. *The Journal of Industrial Economics, XL*(1), 9–35.

Gilbert, R. J. (Ed.). (1992). Symposium on compatibility. *The Journal of Industrial Economics, XL* (1), 1–123.

Holler, M. J., Knieps, G., & Niskanen, E. (1997). Standardization in transportation markets: A European perspective. *EURAS Yearbook of Standardization, 1*, 371–390.
Katz, M. L., & Shapiro, C. (1985). Network externalities, competition and compatibility. *American Economic Review, 75*(3), 424–440.
Katz, M. L., & Shapiro, C. (1986). Technology adaption in the presence of network externalities. *Journal of Political Economy, 94*(4), 822–841.
Katz, M. L., & Shapiro, C. (1994). Systems competition and network effects. *Journal of Economic Perspectives, 8*(2), 93–115.
Kindleberger, C. P. (1983). Standards as public, collective and private goods. *Kyklos, 36*(3), 377–396.
Knieps, G. (1995). Standardization: The evolution of institutions versus government intervention. In L. Gerken (Ed.), *Competition among institutions* (pp. 283–296). London: Macmillan.
Knieps, G. (2003). Competition in telecommunications and internet services: A dynamic perspective. In C. E. Barfield, G. Heiduk, & P. J. J. Welfens (Eds.), *Internet, economic growth and globalization – Perspectives on the new economy in Europe, Japan and the US* (pp. 217–227). Berlin: Springer.
Knieps, G. (2006a). Competition in the post-trade markets: A network economic analysis of the securities business. *Journal of Industry, Competition and Trade, 6*(1), 45–60.
Knieps, G. (2006b). Delimiting regulatory needs. In OECD/ECMT Round Table 129, transport services: The limits of (De)regulation, Paris, 7–31.
Knieps, G. (2011). Network neutrality and the evolution of the internet. *International Journal of Management and Network Economics, 2*(1), 24–38.
Knieps, G. (2013a). Competition and the railroads: A European perspective. *Journal of Competition Law and Economics, 9*(1), 153–169.
Knieps, G. (2013b). Renewable energy, efficient electricity networks and sector-specific market power regulation. In F. P. Sioshansi (Ed.), *Evolution of global electricity markets: New paradigms, new challenges, new approaches* (pp. 147–168). Amsterdam: Elsevier.
Knieps, G. (2013c). *The evolution of the generalized differentiated services architecture and the changing role of the internet engineering task force*. Available at SSRN. Retrieved May 20, 2014, from http://papers.ssrn.com/sol3/papers.cfm?abstract_id=2310693
Knieps, G., Müller, J., & von Weizsäcker, C. C. (1982). Telecommunications policy in West Germany and challenges from technical and market developments. *Journal of Economics* (Suppl. 2), 205–222.
Liebowitz, S. J., & Margolis, S. E. (1990). The fable of the keys. *Journal of Law and Economics, 33*(1), 1–25.
Matutes, C., & Regibeau, P. (1988). Mix and match: Product compatibility without network externalities. *Rand Journal of Economics, 19*(2), 221–234.
Oliveira, R. (2006). *The power of COBOL, for system developers of the 21st century*. BookSurge, North Charleston, SC.
Oren, S. S., & Smith, S. A. (1981). Critical mass and tariff structure in electronic communications markets. *Bell Journal of Economics, 12*(2), 467–487.
Reid, J. (2008). The new features of Fortran 2008. *ACM Fortran Forum, 27*(2), 8–21.
Rohlfs, J. (1974). A theory of interdependent demand for a communications service. *The Bell Journal of Economics and Management Science, IX*(44), 16–37.
Shapiro, C., & Varian, H. R. (1999a). *Information rules: A strategic guide to the network economy*. Boston, MA: Harvard Business School.
Shapiro, C., & Varian, H. R. (1999b). The art of standard war. *California Management Review, 41*(2), 8–32.
Simcoe, T. (2012). Standard setting committees: Consensus governance for shared technology platforms. *American Economic Review, 102*(1), 305–336.
Witt, U. (1991). Reflections on the present state of evolutionary economic theory. In G. M. Hodgson & E. Screpanti (Eds.), *Rethinking economics* (pp. 83–102). Aldershot: Edward Elgar.

Universal Service 7

7.1 Comprehensive Network Opening and Universal Service Objectives

Universal service includes the obligation on the part of the network operator to provide specific services in the whole area for which the obligation is relevant. In particular, this means general access to the service at politically desirable rates, and guaranteeing a certain minimum quality. In traditional regulatory policy universal service obligations in combination with legal entry barriers as well as market power regulation were applied (cf. Kahn, 1970, 1971).

While the introduction, or the maintenance, of legal barriers to entry was meant to avoid inefficient cost duplication and guarantee nation-wide provision of services at socially desired tariffs, the instruments of market power regulation had the objective of at least restricting the exploitation of consumers by excessive prices.

The comprehensive market opening in the network sectors made the application of a disaggregated regulatory policy inevitable; this policy includes:

- complete abolishment of all legal entry barriers,
- disaggregated regulation of market power (cf. Chap. 8),
- universal service funds, insofar as non-profitable universal services are politically desired.

The regulation of network-specific market power must be implemented without regard to the competitive provision of non-profitable universal services. In particular, market power regulation must not be watered down by asymmetric universal service obligations.

7.1.1 Services of General Economic Interest

The term "services of general economic interest" is used in Article 14 and Article 106 (2) in the Treaty on the Functioning of the European Union (originally Treaty of Rome 1957).[1] Although neither these treaties nor the laws derived from them provide a precise definition of the term "services of general economic interest", there is general agreement in the European Community's legal practice that it refers to economic activities relating to public welfare obligations. Moreover, in Article 106 (2) the necessity of applying the rules on competition as far as possible also in the provision of services of general economic interest has been pointed out. Network services in the transport, postal, energy and telecommunications sectors are specifically mentioned. However, other activities undertaken in the common interest might also be relevant (cf. European Commission, 2004, Annex 1). The term "services of general economic interest" particularly comprises the concept of universal service. This concept refers to

> a set of general interest requirements ensuring that certain services are made available at a specified quality to all consumers and users throughout the territory of a Member State, independently of geographical location, and, in the light of specific national conditions, at an affordable price.[2] (European Commission, 2003, Article 50).

After the opening of the network sectors universal services remain an important political objective. Thus an amendment of the German Constitution of August 30, 1994 mandates that Germany's federal government must guarantee, through its regulatory activity, the provision of nation-wide adequate and sufficient services in the competitive telecommunications sector. In addition, the maintenance of universal services remains one of the essential EU political objectives.

7.1.2 Defining the Scope of Non-profitable Universal Services

The demand for universal services initially raises more questions than it provides clear political guidelines. These questions are, in particular: Which services should be provided universally as services of general economic interest? Which quality of universal services should be provided? Is a lowering of quality at the margins of an area of universal service provision acceptable or not? At what rates should universal services be offered?

Answering these questions requires political decisions. There are winners and losers, and in this sense there are no pareto-superior solutions with regard to

---

[1] Consolidated Versions of the Treaty on European Union and the Treaty on the Functioning of the European Union, Official Journal of the European Union, C 83/3, 30.3.2010.

[2] The footnote in this quotation reads as follows: "Cf. Article 3(1) of Directive 2002/22/EC of the European Parliament and of the Council of 7 March 2002 on universal service and users' rights relating to electronic communications networks and services (Universal Service Directive), OJ L 108, 24.4.2002, p. 51".

services, availability and prices. A society's view as to which services should be subsidised is realised through the political process and may vary considerably over time. Thus in the telecommunications sector, providing connection to the narrow-band telephone network or supplying payphones was for a long time considered adequate universal service. In the age of ubiquitous cell phones, payphones have become much less important. Instead, nation-wide supply of broadband Internet access is more and more being debated. In the public transport sector, the focus is still on scheduling and tariff obligations, as well as regular provision of transportation services, which are increasingly questioned because of a lack of public funds.

The question is whether technological change—for instance in the telecommunications sector, because of the development of competing broadband network alternatives—or changed consumer habits will in the future lead to an expansion of the scope of universal service due to increased quality of universal service. Alternatively, at least in the long run, a phasing out of universal services due to sinking costs for the provision of traditional universal services may be expected. In this context, an increasing variety of the standards for universal service (scope, minimum quality, prices, etc.) in different countries and regions (cf., e.g., Jayakar & Sawhney, 2004) is also conceivable. When determining the scope of non-profitable universal services, the division of labour, for instance between the federal authority, individual federal states, counties and municipalities must also be considered (cf. Blankart, 2003, pp. 13ff.). The role of the European Union is very important as well.

7.2 The Instability of Internal Subsidisation Under Competition

In network sectors with legal entry barriers fulfilling universal service objectives was typically associated with internal subsidisation; however, the latter becomes unstable with free market entry.[3]

Let R_i denote the revenue from the ith service of the whole project. A revenue vector $R = (R_1, \ldots, R_n)$ fulfils the cost covering constraint, if:

$$\sum_{i=1}^{n} R_i = C(N), \quad R_i \geq 0 \quad i = 1, \ldots, n \tag{7.1}$$

A revenue vector $R = (R_1, \ldots, R_n)$ fulfils the incremental cost test, if:

$$\sum_{i \in S} R_i \geq \overline{C}(S) \quad \forall S \subset N, \tag{7.2}$$

where $\overline{C}(S) := C(N) - C(N - S)$ denotes the incremental cost of service bundle S, if all other services $N - S$ are provided anyway (cf. Sect. 2.2.2). Thus

---

[3] For a more detailed explanation see Faulhaber (1975), Knieps (1987, pp. 272ff.).

the revenue from each service bundle S must at least cover its incremental cost in order to be non-subsidised. The incremental cost includes the variable cost which is directly attributable to the individual services in bundle S, as well as the additional fixed costs which are required for providing the additional services in bundle S.

A revenue vector $R = (R_1, \ldots, R_n)$ that fulfills the cost covering constraint (7.1) as well as the incremental cost test (7.2), is free of internal subsidisation.

If both the cost covering constraint (7.1) and the incremental cost test (7.2) are fulfilled, it follows that the stability condition is also fulfilled:

$$\sum_{i \in S} R_i \leq C(S) \quad \forall S \subset N \tag{7.3}$$

If there is free market entry, internal subsidisation is unstable in the long run. If the incremental cost test is not fulfilled, then from the cost covering constraint follows:

$$\sum_{i \in S} R_i < \overline{C}(S) \tag{7.4}$$

This constitutes an implicit taxation of the goods $N - S$, because they not only have to cover the firm-specific common costs and their product-group specific common costs, but also share in the incremental cost of service bundle S.

Problems of stability are caused by the uniform price that is frequently demanded with a universal service. Cross-subsidisation, for instance between profitable and non-profitable geographical areas in the interest of tariff unity, can no longer be sustained with free market entry. Free market entry to the lucrative parts of the market ("cherry-picking") would eliminate the surplus which is required for cross-subsidisation of the non-profitable parts.

This is illustrated by Fig. 7.1 (cf. Blankart & Knieps, 1994, p. 241). The supply of network services, for instance postal delivery, as a rule involves considerable economies of scale and scope. The closer the individual households are situated to each other, the lower the long-run average incremental cost $\overline{AC}$ of service provision.[4] Conversely, the cost increases with decreasing population density. In the periphery higher long-run average incremental cost $\overline{AC}$ are to be expected than in the center. The price of provision will thus increase from center Z towards the periphery.

Therefore a politically desired uniform price p_1, cost-covering for the entire system, would not be stable. It would be undercut by cherry-pickers in the profitable centre. Neither would a higher price p_2 leading to cost recovery in the periphery be stable. In contrast, a price that was cost-covering in Z could not be undercut, but it could not be sustained either, because it would not be cost-covering in the peripheral area. Only prices varying over space, i.e. prices that are higher in the periphery

---

[4] These are economies of scale and scope, dependent on the population density of the relevant area.

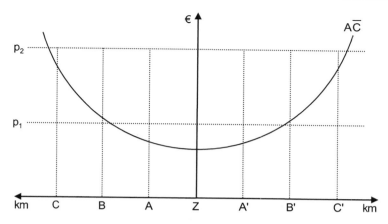

Fig. 7.1 Markets of decreasing density

than in the centre, make stable provision for the whole relevant area possible. Uniform prices like p_1 or p_2, however, include internal subsidisation. User groups located for example between Z and A, or between Z and A', respectively, could break away and provide the services independently, at a lower price. Therefore these prices are not sustainable in case of free market access.

Just like this internal subsidisation between densely and sparsely populated areas, cross-subsidisation, where the surplus from profitable services is used to subsidise other services (all over the country or only in the periphery), is not sustainable either. For example, for a long time telephone services were cross-subsidising directory assistance and in the postal services letters were cross-subsidising parcels. These forms of internal subsidisation are not stable under competition either.

7.3 The Concept of the Universal Service Fund

In network sectors with free market entry universal service objectives can no longer be financed by means of internal subsidisation (cross subsidisation) and there is no longer only one possible network provider that can supply them. Thus it becomes necessary to realise politically desirable universal service objectives without favouring, or discriminating against, individual market players. A transparent way of solving this problem is a universal service fund, which will be characterised in the following.

The basic concept of the universal service fund consists of the creation of a level playing field for all active and potential suppliers, both in the profitable and the non-profitable markets. In doing so one must differentiate between the revenue side and the expenses side of the fund.

Traditional cross subsidisation is an implicit taxation of profitable services. The universal service fund is also based on intra-sector financing, insofar as all providers of profitable services in the relevant sector are in equal measure drawn upon to finance the fund. All suppliers of profitable services pay a sales-dependent universal service tax into the fund, so that there will be no discrimination against, or favouring of, individual suppliers because of firm size. These revenues cover the subsidy requirements for providing non-profitable universal services.[5]

The expenditures side of the universal service fund is organised as follows: Non-profitable universal services are supplied at the lowest possible cost of subsidy, which is financed out of the universal service fund (cf., e.g., Blankart & Knieps, 1989, pp. 593f.). If an individual universal service is in fact profitable, or if, due to technological progress, subsidisation is no longer necessary, then subsidisation from the universal service fund must be stopped. The lowest cost supplier of non-profitable universal services in a specific sector can be determined in the context of a public tender (cf. Sect. 5.3). Auction mechanisms were developed, for example, for universal services in the telecommunications sector, in order to determine the supplier and the extent of the necessary subsidy (cf., e.g., Sorana, 2000; Weller, 1999). Regarding the postal sector, the potential of auctions for ensuring postal universal services has also been studied (cf. Borrmann, 2004).

The strength of the universal service fund solution is that the number of sub-networks requiring subsidisation is determined endogenously, i.e. during the process itself (and not predetermined by a regulatory agency). Only the politically desired uniform prices (or price ceilings, respectively) and the minimum quality at which universal services have to be provided are specified in advance. All the same, an excessive fragmentation of the supply of universal services into different suppliers is not to be expected, because the economies of scope of jointly serving adjacent areas will be exhausted under competition.

Under competition, without politically determined universal service objectives, prices would vary automatically, according to the differing costs of providing services in different areas. However, affordable prices do not necessarily have to be uniform prices. An alternative possibility would be price ceilings for the provision of universal services. In the past, uniform prices served the purpose of leading to substantial profits in lucrative sub-areas, which could then be used for cross-subsidising non-profitable sub-areas. However, this process could only work as long as there were legal barriers to entry. After the comprehensive opening of the markets, the revenue for financing the deficit has to be earned in a way that is compatible with market entry. After market opening, uniform prices become unstable, because they can be selectively undercut by competitors. A positive effect of competition is that incentives for cost-cutting are created, which can be expected to lead to a decreasing need for subsidisation in the universal service sector. Everything considered, the introduction of price ceilings will thus result in smaller subsidy requirements.

---

[5] Alternatively, financing from the national budget is also a conceivable possibility.

7.4 Universal Services in Telecommunications Markets

Universal services have traditionally been of great importance in telecommunications markets. In the past access to narrowband local telephone networks at uniform tariffs could only be maintained by means of internal subsidisation und constituted the central universal service. Further examples of universal services are access to directory assistance, the provision of public payphones, unobstructed and free of charge access to emergency call services or the provision of directories listing the participants of public voice telephony services.

Due to the enormous technological progress in communications networks and the Internet, communications markets are undergoing a massive upheaval. Examples for this are interactive broadband communications networks, Internet-based telephony, mobile communications networks, etc. This gives rise to the question if and to what extent the traditional concept of universal service is still valid, and which reforms are conceivable in the future.

Even after the comprehensive market opening in the telecommunications sector that has been taking place worldwide, the political objective of ensuring a minimum supply of telecommunications services at politically desired prices remains beyond dispute. However, the necessary scope of these telecommunications services is still under debate. Unlike various other network sectors, the telecommunications sector is highly dynamic, and thus the problem naturally arises whether and to what extent innovative communications services should also be regarded as part of universal service.

The Universal Service Directive of 7 March 2002 defined the scope of universal service narrowly.[6] The requirement of connection to the public telephone network at a fixed point and at an affordable price is limited to narrowband access and does not include ISDN. In most European countries telecommunications networks are constructed in such a way that they also enable narrowband Internet access. However, no fixed transfer rate was specified for the European Union, so that the potentials of wireless technology can also be utilised for the provision of universal services. In certain cases where connection to the public telephone network is definitely not sufficient for satisfactory Internet access, member states should have the option of prescribing an upgrade of the connection, so that transfer rates sufficient for Internet access are supported. The necessary cost of this upgrade could be apportioned to the cost of the universal service obligation. Further components of the universal service obligation are member directories, directory assistance, as well as a sufficient number of public coin-operated and card-operated payphones. In addition, there are special provisions for handicapped users and users with special social needs.

---

[6] Directive 2002/22/EC of the European Parliament and of the Council of 7 March 2002 on universal service and users' rights relating to electronic communications networks and services (Universal Service Directive), OJ L 108, 24.4.2002, pp. 51ff.

This restrictive definition of the scope of universal service has not been undisputed. This is shown, for example, in the extensive controversy about the US Federal Communications Commission's practice of defining universal service programmes.[7] In its interpretation of the relevant Section 254 of the Communications Act of 1996 the FCC did not limit itself to network access for cost-intensive rural areas and for low-income households, but also included the provision of schools, libraries, and certain health services in universal service programs. All schools and libraries receive substantial subsidisation of Internet access and telecommunications services. The providers of health services in rural areas receive, among other things, a subsidisation of broadband telecommunications services. In the following, this expansion of universal service was criticised, one of the arguments against it being that it was not the FCC, as the regulatory agency, but the legislator who had the competence to decide upon such an expansion of universal service.[8]

A periodical review of the necessary scope of universal service, on the part of the commission, in accordance with the EU Universal Service Directive, refers both to the question whether traditional universal services (e.g. public coin-operated and card-operated payphones, member directories and directory assistance) should still be part of universal service, and the question whether the scope of universal service should be expanded to include mobile and broadband communications services. For a long time, the traditional specification of the scope of universal service has been adhered to, because competition in the mobile telecommunications market makes affordable access to mobile services possible, while broadband Internet access has not been considered as meeting the criterion of use of the service by a "majority of consumers" (European Commission, 2006, pp. 2f.). Not only in the EU member states, but worldwide the traditional narrow definition of the scope of universal service has long been applied, in particular access to narrowband voice telephony services at affordable prices. However, the growing importance of Internet telephony (Voice over IP telephony) and other time sensitive Internet applications leads to a challenge of the traditional narrowband based universal service concept (cf. e.g. Xavier, 2006; Jayakar & Sawhney, 2004). As a consequence, on the level of broadband Internet traffic services universal service objectives may gain momentum and subsequent subsidies for the socially desired premium traffic services may become necessary (cf. Knieps, 2011, pp. 8ff.). The developmental dynamics of the telecommunications sector affect not only the specification of the possible scope of universal services, but also the demarcation of services on which a universal service tax should be imposed. It is to be expected that the spectrum of profitable networks and services will continually change under competition. In order to raise a universal

---

[7] CC Docket No. 96-45, In the Matter of Federal-State Joint Board on Universal Service, May 8, 1997; April 10, 1998; Proposed Third Quarter 2001, Universal Service Contribution Factor (CC Docket No. 96-45, June 8, 2001).

[8] CC Docket No. 96-45, April 10, 1998, Congressional Intent Regarding Federal Universal Service Programs, Statement of Sen. Byron Dorgan, pp. 134–155.

service tax dependent on sale, however, it is sufficient to undertake a negative demarcation of those having to pay the tax. But high market shares are not a suitable criterion for this—instead, it should be permitted to make all suppliers of communications services contribute to this tax. In practice, however, problems with separating taxable from tax-exempt added value are unavoidable, in particular with regard to data transfer and data processing. This is shown very strikingly in the US telecommunications sector, in the extensive debate in CC Docket 96-45 concerning the question if sales from Internet services can be drawn upon for the financing of universal service programs. In a 2006 ruling the FCC decided that in the future the universal service tax will also be imposed on mobile communications and Internet telephony.[9]

7.5 Universal Services in Postal Markets

In the past the major argument against the liberalisation of postal markets has been that "cherry-picking" activities by entrants would make the traditional cross-subsidisation unstable. Just like cross-subsidisation between densely and sparsely populated areas, cross-subsidisation, where the surplus from profitable services is used to subsidise other services (all over the country or only in the periphery), is not sustainable either. For example, in the past letters were cross-subsidising parcels. These forms of cross-subsidisation are not stable under competition. Therefore, the introduction of competition would conflict with the traditional universal service objectives. Thus, the liberalisation of European postal markets was a time consuming process, abolishing legal entry barriers only gradually due to universal service arguments (cf. Knieps, Zenhäusern, & Jaag, 2009, pp. 89f.). Free entry into all postal markets including letter services has become a reality in all EU member countries since January 2013.

In liberalised postal markets, universal services remain a worldwide important political objective, focussing on the provision of nation-wide adequate and sufficient services. In the following we shall see that it is in fact possible to make competition in postal markets and politically desired universal service objectives compatible.

In liberalised postal markets universal service objectives can no longer be financed by means of internal subsidisation and there is no longer only one possible service provider that can supply them. In closed network sectors fulfilling universal service objectives was typically associated with internal subsidisation; however, the latter becomes unstable with free market entry. The term "cherry-picking" by market entrants is strongly related to cross-subsidisation. If one is interested in the relevance of such cross-subsidisations in the postal markets, one immediately runs into a serious problem of terminology. In the past, legally protected postal

---

[9] WC Docket No. 06-122, In the Matter of Universal Service Contribution Methodology, June 27, 2006, p. 3.

monopolies considered services cross-subsidised if their prices were below average costs, based on arbitrarily allocated overhead costs (like the relative usage time, etc.). In particular, this left excessive strategic room for deriving politically accepted cost/benefit ratios for socially desired services. As has already been known for a long time from the study by Clark (1923) on overhead costs, there is no economic reason to do so. Instead, from an economic point of view, the question arises as to whether the allocation of costs is acceptable in such a way that no incentive is created to separate from an efficient joint production. This immediately leads to the incremental cost tests, developed by Alexander (1887) in the context of railway economics and extended by Faulhaber (1975) with game theoretical tools. According to the incremental cost test no cross-subsidisation occurs, if the revenue of each product (or coalition of products) at least contributes its incremental costs. In other words, overhead costs should not play any role as criteria for cross-subsidisation, because their specific allocation does not influence the incentive to give up joint production.

Thus it becomes necessary to realise politically desirable universal service objectives without favouring, or discriminating against, individual market players. A transparent and symmetric way of solving this problem is the implementation of a universal service fund, if the universal service objectives cannot be provided spontaneously via the market without subsidies. According to the basic principles of a universal service fund, profitable postal services have to pay an entry tax and loss-making services are provided by the winner of an auction requiring the lowest amount of subsidies.

Under competition, without politically determined universal service objectives, prices would vary automatically, according to the differing costs of providing services in different areas. However, affordable prices do not necessarily have to be uniform prices. An alternative possibility would be price ceilings for the provision of universal services. In the past, uniform prices served the purpose of leading to substantial profits in profitable sub-areas, which could then be used for cross-subsidising non-profitable sub-areas. However, this process could only work as long as there were legal barriers to entry. After the liberalisation of the postal markets, the revenue for financing the deficit has to be earned in a way that is compatible with market entry. After market opening, uniform prices become unstable, because they can be selectively undercut by competitors applying "cherry-picking" strategies. A positive effect of competition is that incentives for cost-cutting are created, which can be expected to lead to a decreasing need for subsidisation of the politically desired universal postal services.

7.6 Questions

7-1: Internal Subsidisation
Explain why internal subsidisation is incompatible with free market entry.

7-2: Universal Service Fund
Explain the basic elements of a universal service fund.

7-3: Case Study Telecommunications
What are the specific characteristics of the telecommunications sector which have to be taken into account when determining the extent of universal service.

References

Alexander, E. P. (1887). *Railway practice*. New York: G.P. Putnam's Sons.
Blankart, Ch. B. (2003). Universaldienst und Liberalisierung: Die föderale Dimension - Konsequenzen für das neue TKG -, *Telekommunikations- & Medienrecht*. TKMR-Tagungsband, pp. 13–17.
Blankart, C. B., & Knieps, G. (1989). What can we learn from comparative institutional analysis? The case of telecommunications. *Kyklos, 42*, 579–598.
Blankart, C. B., & Knieps, G. (1994). Das Konzept der Universaldienste im Bereich der Telekommunikation. *Jahrbuch für Neue Politische Ökonomie, 13*, 238–253.
Borrmann, J. (2004). Franchise bidding for postal services in rural regions. *The B.E. Journal of Economic Analysis and Policy, 4*(1), 10.
Clark, J. M. (1923). *Studies in the economics of overhead costs*. Chicago: The University of Chicago Press.
European Commission (2003). *Green paper on services of general interest*. Brussels, 21 May 2003, COM(2003) 270 final.
European Commission (2004). *White Paper on services of general interest*. Brussels, 12 May 2004, COM(2004) 374 final.
European Commission (2006). *Report regarding the outcome of the Review of the Scope of Universal Service in accordance with Article 15(2) of Directive 2002/22/EC*, Brussels, April 7, 2006, COM(2006) 163 final.
Faulhaber, G. R. (1975). Cross subsidization: Pricing in public enterprises. *American Economic Review, 65*, 966–977.
Jayakar, K. P., & Sawhney, H. (2004). Universal service: Beyond established practice to possibility space. *Telecommunications Policy, 28*, 339–357.
Kahn, A. E. (1970). *The economics of regulations: Principles and institutions* (Economic principles, Vol. 1). New York: Wiley.
Kahn, A. E. (1971). *The economics of regulation: Principles and institutions* (Institutional issues, Vol. 2). New York: Wiley.
Knieps, G. (1987). Zur Problematik der internen Subventionierung in öffentlichen Unternehmen. *Finanzarchiv, N.F., 45*, 268–283.
Knieps, G. (2011). Market driven network neutrality and the fallacies of internet traffic quality regulation. *International Telecommunications Policy Review, 18*(3), 1–22.
Knieps, G., Zenhäusern, P., & Jaag, C. (2009). Wettbewerb und Universaldienst in europäischen Postmärkten. In G. Knieps & H.-J. Weiß (Eds.), *Fallstudien zur Netzökonomie* (pp. 87–109). Wiesbaden: Gabler.
Sorana, V. (2000). Auctions for universal service subsidies. *Journal of Regulatory Economics, 18*(1), 33–58.
Weller, D. (1999). Auctions for universal service obligations. *Telecommunications Policy, 23*, 645–674.
Xavier, P. (2006). *What rules for universal service in an IP-enabled NGN Environment*. Invited Paper Presented to an ITU Workshop on "What rules for an IP-enabled NGN?", 23–24 March, Geneva.

Market Power Regulation

8.1 Localisation of Network-Specific Market Power

8.1.1 Competition Versus Market Power

Since the comprehensive abolishment of legal barriers to entry in (almost) all network sectors network economics has undergone a paradigm shift. Whereas before the opening of the markets the controversial question was if and to what degree competition in network industries could function at all, in the meantime the central controversy of network economics has shifted to the division of labour between sector-specific regulation and general competition law.

From the perspective of economic order the application of sector-specific regulatory intervention constitutes a massive intervention in the market process and thus always requires a particularly well-founded justification. That the antitrust provisions of general competition law should be applied to the opened network sectors, too, is beyond dispute. Sector-specific regulatory interventions with competition policy objectives, on the other hand, are only justified if there is network-specific market power.[1] Insofar as vague legal terms originating from general competition law—such as, for instance, market dominance—are being used to determine the need for sector-specific intervention, they have to be corroborated by a localisation of market power that is substantiated by network economics.

---

[1] Technical regulatory functions (network security, allocation of frequencies, telephone number administration, the definition and enforcement of geographical limits of air traffic control jurisdictions etc.) and the pursuit of universal service objectives by means of entry-compatible instruments (e.g. universal service fund) also constitute long-term sector-specific regulatory tasks, but should not be confused with market power regulation. However, technical regulation may also fulfil the role of a precondition for the functioning of competition. For example, competition on the markets for international roaming could not evolve, due to the absence of adequate technical regulations (cf. Knieps & Zenhäusern, 2014).

A suitable economic reference model for establishing the regulatory activity necessary for disciplining market power in network sectors must be able to take into account the essential characteristics of networks (economies of scale and scope, network externalities, network variety) without automatically equating them with market power. A basic distinction has to be made between competition between networks (intermodal competition, platform competition) and competition on the different network levels (cf. Sect. 1.2).

The following quotation characterises the concept of market power established in antitrust literature:

> The term "market power" refers to the ability of a firm (or a group of firms, acting jointly) to raise price above the competitive level without losing so many sales so rapidly that the price increase is unprofitable and must be rescinded (Landes & Posner, 1981, p. 937).

Stable market power means that the long-term characteristics of the market being considered (in particular production conditions and demand conditions) permit stable economic profits which will not be eliminated by competition (e.g. arbitrage activities). This has to be distinguished from short-term economic profits which may occur due to temporary characteristics of the market under consideration, but which will in the following be rapidly eliminated by the competition from other suppliers.

8.1.2 Market Power Due to Economies of Scale?

The "invisible hand" of perfect competition, as formulated by general equilibrium theory, disregards the existence of economies of scale altogether (cf. Walras 1874/1877). Atomistic competition with a large number of suppliers who take the price as given and have no other parameters of strategy at their disposal either, does thus not constitute a relevant competition concept in network sectors. Instead, competition-theoretical concepts in network economics have to take economies of scale and scope into consideration for model analysis, because these are central characteristics for network sectors. Thus, the many faces of competition have to be considered, taking into account the role of economies of scale, product differentiation and the evolutionary search for new products and production technologies.

In the past, the method for "proving" market power in antitrust cases frequently consisted of, in the first place, defining the relevant market in which the market share of the indicted firm was to be determined; secondly, calculating this market share; and thirdly, deciding whether this market share was sufficiently large to permit a reliable conclusion regarding the existence of market power. This method is not suitable for determining stable market power and the necessary regulatory policy in network industries. Market shares do not constitute a reliable criterion for the existence of market power and the resultant high (non-competitive) prices. In fact, high market shares can be the result of low prices. In particular, the role of potential competition has to be taken into consideration.[2] From the perspective of modern industrial economics, too, no causal relation between the existence of economies of scale and market power can be derived (cf. e.g. Schmalensee, 1989, pp. 951ff.).

8.1.3 Network-Specific Market Power

Due to comprehensive market opening network industries such as railway transport, air traffic, telecommunications and energy have lost their status as special sectors exempted from general competition law. Therefore the demarcation between the specific tasks of sector-specific regulation versus general competition law needs to be redefined, so that potentials for competition can be comprehensively utilised in opened network sectors also. End-to-end market power regulation in legally protected network monopolies has to be replaced by disaggregated market power regulation.

Stable network-specific market power can only be proven to exist in case of a combination of a natural monopoly and irreversible costs, that is, if there is a monopolistic bottleneck (cf. Knieps, 1997, pp. 327ff., 2011a). The conditions for a monopolistic bottleneck facility are fulfilled

- if a facility is necessary for reaching customers, i.e. if no alternative facility serving as an active substitute exists. This is the case when, based on economies of scale and economies of scope, a natural monopoly does exist so that a single provider is able to provide the capacities of a facility more cheaply than several providers;
- and if, at the same time, no potential substitute is available. This is the case when the costs of the facility are irreversible and the facility cannot reasonably be duplicated as a way of disciplining the active provider.

The owner of such a monopolistic bottleneck has stable market power, even if all market participants are perfectly informed, all consumers are prepared to switch providers, and minor price changes lead to a shift in demand. Thus network-specific market power on the part of the established firm can only be expected in those areas that are characterised not only by a natural monopoly, but at the same time also by irreversible costs. Irreversible costs are no longer decision-relevant for the established firm—in contrast to the potential competitor, who is faced with the decision whether to invest in a given market or not. Thus the incumbent has lower decision-relevant costs than the potential competitors. This enables strategic behaviour, so that inefficient production or economic profits no longer necessarily result in newcomers entering the market.

---

[2] On the preconditions for functioning potential competition as a substitute of active competition, cf. Demsetz (1968, p. 58).

In the absence of irreversible costs, however, as a result of the disciplining effect of potential competition, a natural monopoly does not lead to stable market power. This is true regardless of the size of the relevant network operator's market share, because inefficient providers of services that fail to meet market needs will be replaced by new entrants, owing to the pressure of potential competition. In this case there is no need for regulatory action to limit the active operator's control over the market. The monopolistic bottleneck theory does not attempt to deny the information problems encountered to varying degrees on real markets. Ex ante stable market power cannot be deduced from the existence of information problems, however, because markets tend to be efficient at (endogenously) developing institutions to overcome their information problems.

Switching costs do not cause monopolistic bottleneck situations either. They occur in many areas of the economy, for instance monthly or annual season tickets for concerts that cannot be transferred if the holder moves house, or the costs incurred by firms when employees leave as soon as they have learnt the ropes, etc. Switching costs are no justification for regulatory measures and can be left to the market's own problem-solving ability (cf. e.g. von Weizsäcker, 1984; Tirole, 1989, Chap. 8).

However, the existence of network externalities is no justification for sector-specific regulation either. The essential feature of such externalities is that for an individual the advantage of being part of a network depends not only on its technical specifications—its standard—but also on how many others are part of it (cf. Sect. 6.1.1). Where there are positive network externalities, the benefit for the individual increases with the number of other network members, in other words, the number of people using the same standard. In the absence of network-specific market power, negotiations between network operators are efficient, because both sides stand to benefit from the agreements. Thus for instance a conflict between the utilisation of the network externalities of a common user base and a preference for differing standards in telecommunications networks is unlikely. In contrast, ensuring access to monopolistic bottlenecks constitutes a regulatory task, because network-specific market power enables strategic behaviour which also obstructs the full utilisation of the positive externalities of network access (cf. e.g. Blankart & Knieps, 1995).

Thus the objective of the theory of monopolistic bottlenecks is to tackle the specific remaining regulatory need of ex ante stable derivable network-specific market power. In contrast, for network areas with economies of scale and scope, but without irreversible costs—as for all other markets—the provisions against collusion and the abuse of a dominant position as laid down in general competition law are sufficient. However, as on all other markets, the burden of proof whether market power exists and whether it is being abused (cf. for instance § 19 GWB or Article 101, 102 Treaty of Lisbon[3]) rests with the competition authorities. In contrast to ex

---

[3] Consolidated Versions of the Treaty on European Union and the Treaty on the Functioning of the European Union (2010/C 83/01), Official Journal of the European Union, 30.3.2010, Title VII.

8.2 Disaggregated Identification of Competitive Potentials in Network Industries

Table 8.1 Competition versus network-specific market power

Network areas	With irreversible costs	Without irreversible costs
Natural monopoly	Monopolistic bottleneck	Potential competition
No natural monopoly	Active competition	Active competition

ante regulation, such interventions in the market process should only be made case-by-case and ex post. In doing so, competition authorities have to weigh two possible sources of error against each other. A false positive error occurs, when the competition authority intervenes in the competitive process, even though competition is functioning and there is no need for any active competition policy measures. A false negative error occurs when the competition authority fails to act, even though competition policy measures are indeed called for.[4]

The conclusion to be drawn from this is that network-specific market power only occurs, when neither active nor potential competition exists. Competition in networks thus means active competition between different network infrastructure providers as well as active and potential competition between suppliers of network services. The localisation of network specific market power can be summarised in Table 8.1:

8.2 Disaggregated Identification of Competitive Potentials in Network Industries

The potentials of active and potential competition have to be localised in networks in a disaggregate manner. In order to achieve this it is useful to differentiate the markets for network services from the upstream markets for infrastructure capacities (cf. Sect. 1.2).

8.2.1 Competition on the Network Service Level

8.2.1.1 Traffic Services

Active and potential competition is functioning on transport markets. As long as transport firms can freely enter the market, even the supply of transport services in the form of a network and its concomitant economies of scale and scope do not imply monopoly power, because if a firm has high profits, competitors will at once appear. There is no credible threat potential to keep rival firms from entering the market, because on the level of transport services decision-relevant costs are comparable for both the incumbent and the potential competitor. Thus for supplying transport on a rail network, for instance, cost irreversibilities are not significant. The

---

[4] In the context of competition policy regarding predatory pricing the comparison of these error types has been analyzed by Joskow and Klevorick (1979).

deployment of railway trains is not tied to specific tracks; just like airplanes and trucks, they are geographically mobile.

However, a precondition for functioning competition is that every (active and potential) supplier of transport services is granted equal access conditions to transport infrastructures. As long as incumbents have privileged access to scarce infrastructure capacities, they enjoy unjustified competitive advantages which may lead to the emergence of market power in transport markets that would otherwise be competitive.

While the theory of contestable markets analyses exclusively the role of potential competition with identical cost functions for both the active supplier and the potential competitor (cf. Baumol, 1982; Panzar & Willig, 1977), competition in markets for transport services is by no means limited to potential competition. A newcomer may enter the market with no intention of duplicating the established firm. The important entrepreneurial challenge is to face the potentials for active competition, achieved by means of technological and product differentiation and innovations (product and process innovations).

For railway traffic, too, active competition on busy tracks can be expected to lead to efficient tariff offers, such as, for instance, increasing incentives for cost efficiency and the pressure to offer demand-compatible transport services. For passenger transport, competitive pressure reveals whether the length of the supplied trains and their frequency are compatible with demand. Traffic concepts that were formerly administratively specified (for example synchronised timetables) are being challenged, if the customers in the market do not reward them, that is, if there is no analogous demand. Regular tours of (almost) empty "ghost trains" cannot be kept up under competition. On the other hand, incentives emerge for supplying flexible additional transport offers in peak times. The market entry of new transport firms leads to a considerable extension of the range of services offered, as well as to a broadening of choices regarding price and transport quality. Part of this process is finding and taking advantage of market niches, such as establishing an express service for goods and people by means of developing a heavy-duty computerised logistics system. But improvements of service are also possible for short-distance traffic, for instance by offering a denser time-table with optimised connections. Thus, in addition to the pressure of potential competition, active competition between different transport firms also has a potential that should not be underestimated (cf. Knieps, 2006b, pp. 202ff.).

8.2.1.2 Telecommunications Services

Some markets for telecommunications services on long-distance networks are still characterised by economies of scale and scope. Nevertheless, there is functioning competition in such telecommunications services networks. Inefficient suppliers will be replaced by cheaper ones, when free market entry is possible. Even if the incumbent's initial market share is high, it will decrease rapidly and considerably, if production is inefficient or services are not market-compatible, because customers are not tied to a specific supplier and can react immediately to price reductions in the market (cf. Knieps, 2006c, pp. 152f.).

8.2 Disaggregated Identification of Competitive Potentials in Network Industries

On competitive telecommunications markets service competition and infrastructure competition must be granted the same opportunities and must not be distorted by regulatory measures. Regulatory constraints motivated by short-term considerations that obstruct infrastructure competition unilaterally suffer from the serious disadvantage of setting the wrong incentives long-term. On the one hand they create insufficient incentives for market entrants to make profitable infrastructure investments themselves, on the other hand regulatory requirements to offer services below costs always lead to discrimination against the incumbent network operator, as nobody would willingly offer network components under such conditions. On service markets offered on the basis of telecommunications infrastructures there is at present vigorous innovation competition. Nevertheless, there is an ongoing "network neutrality" debate on the role of the regulation of Internet traffic management (cf. Knieps, 2011b).

8.2.2 Competition on the Infrastructure Management Level

In railway and air traffic continuous control and coordination of traffic movements is indispensible. In order to achieve this, railway traffic control and air traffic control, respectively, are necessary; their task is not only to guarantee traffic safety, but also to allocate the existing traffic infrastructure capacities. For road traffic, too, traffic management systems will play an increasingly important role in the future.

It must be taken into consideration that the supply of transport services requires the simultaneous access to a traffic infrastructure and a traffic management system, no matter if these functions are vertically integrated in the hand of one firm or are offered by different firms. Although airport operators, airlines and air traffic control authorities have to cooperate to be able to guarantee smooth-flowing traffic, they have traditionally been both organisationally and institutionally separated.

For railway traffic the situation has been different for a long time, with all functions being integrated in the hand of the respective national railway companies, and, as a rule, only minimal cooperation between the different national railway companies. But here, too, a development towards a disaggregated regulatory approach can be observed. Competition on railway networks is only possible if railway companies are granted unrestricted access both to the tracks and to the services of railway traffic control systems (cf. e.g. Knieps, 2006d, pp. 15ff.). Railway traffic control systems constitute the crucial link between track and service. Both the flow of railway traffic and the implementation of repairs of the tracks have to be coordinated by railway traffic control systems. This coordination effort is, as in the case of air traffic, in principle not dependent on whether one or several railway companies are active on a given track network. Instead, it depends on the number of trains and their speed.

It is obvious that active competition between different suppliers of railway traffic control systems cannot work. One airplane or one individual train must only be controlled by one institution at a time, if chaos and accidents are to be avoided. The control competency must remain in one hand for each given time period. This gives rise to the problem of a "natural" boundary of a regional control area on the one hand and the coordination between different control areas on the other hand.

However, traffic control systems do not have the characteristics of monopolistic bottlenecks. They can be characterised as natural monopolies, whose geographical boundaries have to be clearly defined and enforced by technical regulations. But from this no network-specific market power can be derived, because the computer software necessary for building traffic control systems and the knowledge required are not sunk costs invested into specific geographical locations. While for traffic services competitive pressure is also guaranteed by active hit and run entry, for traffic control systems competitive bidding should be applied. The object of such an auction would be the supply of traffic control for an ex ante given geographical area during a well-defined time period. The contract would be awarded to the bidder able to supply traffic control in a cost-covering manner at the lowest tariff.

The reorientation of the European infrastructure policy towards the objective of building and developing trans-European networks followed by Connecting Europe initiatives (cf. European Commission, 2011) leads directly to the necessity of promoting the interoperability of individual national networks. Considerable coordination efforts are required, in particular in the areas of traffic management and traffic control systems.

The former railway monopoly led to a predominantly national orientation of the capacity management of railway tracks and a time-table design for allocation of track capacities that was based on national interests. Cross-border coordination and cooperation within the International Union of Railways was not a top priority. This applied to both standardisation efforts and coordination and cooperation in track management. Optimisation efforts were focused on national systems. In the meantime, increasing tendencies towards transborder interoperability are emerging, for instance the development and deployment of the European Rail Traffic Management System (ERMTS).[5]

Like air traffic control systems, railway traffic control systems have considerable cross-border potential. Competition on the European railway transport markets and the concomitant increase of demand for European railway transport require a consistent internalisation of cross-border restrictions. For instance, the technical boundaries of a facility (e.g. telecommunications, radio) should no longer be aligned with political boundaries between countries. Cross-border system advantages have to be exhausted thoroughly, so that competition on the European markets for railway transport can fully develop.

The development of an integrated European railway traffic control system would be greatly facilitated by establishing independent railway traffic control agencies, such as exist for air traffic control. As long as there is no such development in the

---

[5] Directive 2008/57/EC of the European Parliament and of the Council of 17 June 2008 on the interoperability of the rail system within the Community (OJ L 191/1, 18.7.2008) and the subsequent Commission Decision of 25 January 2012 on the technical specification for interoperability relating to the control-command and signalling subsystems of the trans-European rail system (OJ L 51/1, 23.2.2012).

direction of an integrated European system, at least the opportunities of an intensive coordination and harmonisation of railway traffic control systems should be comprehensively utilised—e.g. via more vigorous standardisation efforts and time-table coordination (cf. Sect. 6.5).

8.2.3 Competition on the Network Infrastructure Level

Network services cannot be provided without at the same time having access to the complementary network infrastructures. Thus transport services not only require vehicles (e.g. a train or an airplane), but also access to traffic infrastructures (e.g. a railway track or a landing slot). On the network infrastructure level competition potentials are not impossible, either. One has to differentiate between network areas where network competition exists and network areas that are monopolistic bottlenecks.

8.2.3.1 Fixed Line Telecommunications Networks

Since the comprehensive opening of the telecommunications networks in Europe in 1998 there has been massive investment in alternative infrastructures for long-distance networks. For long-distance networks, both active and potential competition by alternative interconnection network providers is assured. A good example is competition in wireless networks—for example mobile telecommunications, satellite or microwave systems—as well as network competition by alternative wired network providers. By now in Germany several competitors operate their own long-distance networks. A multitude of interconnection points are provided by alternative network providers. At present, there are several firms in Europe that are building their own optical fibre backbone infrastructures. These high investments in alternative infrastructures have led to intense competition on the national and transnational markets for transmission capacity. Upstream markets for long-distance telecommunications are therefore characterised by many forms of active and potential competition.

Due to active and potential competition in the construction of alternative long-distance telecommunications network infrastructures established network providers do not have network-specific market power. Competition fulfils the function of disciplining market power. Consequently, it is to be expected that private interconnection bargaining between different network owners will lead to economically efficient solutions and prove to be incentive compatible for all market participants. Strategic behaviour on the basis of network-specific market power can be excluded, because each negotiating party can easily be replaced by an alternative (potential) network operator.

The situation is different, however, for local, wired networks. At the beginning of market opening these networks fulfilled the characteristics of monopolistic bottlenecks. Building them requires high irreversible costs, and duplicating such local networks was not reasonable from an economic perspective. In the meantime, however, a shrinking of such monopolistic bottlenecks can be observed. Since the abolishment of legal barriers to entry, in areas of high population density parallel local network providers enter the market. In the meantime, alternative network access technologies ("wireless local loop", optical fibre etc.) play an increasingly larger role. Thus cable TV providers, for instance, offer more and more interactive communications networks, which, beside cable TV, also include high-bit rate Internet traffic as well as IP-based telephony (triple play). In addition, broadband mobile telecommunications connections (UMTS) are also increasingly being used. Thus a gradual phasing out of access networks (local loop) as monopolistic bottlenecks can already be observed today. This proves the necessity of periodically checking the phasing-out potentials of monopolistic bottlenecks (cf. Knieps & Zenhäusern, 2010). It is necessary to differentiate between those subsets of access networks where a monopolistic bottleneck scenario still exists and those subsets of access networks where—for instance due to wireless local loop and alternative cable network providers—there is already functioning active and/or potential competition.

8.2.3.2 Mobile Telecommunications Networks

An important precondition to enable competition among different carriers is technical regulation in the form of number portability, irrespective whether the communications network is fixed or mobile. Subscribers of publicly available telephone services can only switch from one provider to another if they can retain their numbers independently of the undertaking providing the service. In fixed networks number portability was granted on the national and international telecommunications markets before global entry deregulation was implemented. But after 2002 for national mobile telecommunications, too, number portability was implemented as a technical regulation, and thus irrespective of market power considerations of the carriers involved. However, technical regulation to allow carrier number portability has never been implemented in international markets for roaming leading to a subsequent lack of competition (cf. Knieps & Zenhäusern, 2014).

In the meantime, access and call origination on mobile networks have also been deleted from the list of markets scheduled for possible ex ante regulation (cf. Knieps & Zenhäusern, 2010). End customers can choose between the services offered by the different network providers and those offered by service providers. Even after a customer has chosen a specific mobile telecommunications provider, the latter's position does not turn into the position of a monopolistic bottleneck supplier, because switching costs are declining ever more strongly (for instance with a prepaid card). Irrespective of this development, switching costs do on principle not lead to a monopolistic bottleneck scenario (cf. Sect. 8.1.3).

This leaves the question whether mobile telecommunications networks have to be regarded as monopolistic bottlenecks from the perspective of network interconnection. Looking at the different interconnection options where at least one mobile telecommunications network is involved (mobile telecommunications network-mobile telecommunications network, mobile telecommunications network-fixed-line network, fixed-line network-mobile telecommunications network) the question arises whether termination in a mobile telecommunications network constitutes a monopolistic bottleneck, because a particular mobile telecommunications customer cannot be reached otherwise. However, what is crucial here is the fact that alternative mobile telecommunications network providers exist which can, with adequate means, be chosen as a (partial or complete) substitute. Even though the caller has to pay for the call, the person being called has an incentive to make sure that the caller does not have to pay exorbitant termination charges. For instance, the person being called may ultimately have to pay for the call himself, if it is a family member who calls, or he may fear that, due to the high termination charges, he will get fewer calls or none at all. Thus in particular for network participants with high demand of incoming calls there are incentives to switch to an alternative mobile telecommunications provider with lower termination charges.

In addition, incoming and outgoing calls exist in a close substitution relationship—and thus in competition—with each other. Therefore they have to be regarded as a single market; high fees for incoming calls lead to bait calls from fixed-line networks and high fees for outgoing calls lead to bait calls from mobile telecommunications networks.

8.2.4 Monopolistic Bottlenecks on the Network Infrastructure Level

Network-specific market power exists in those network areas that have the character of monopolistic bottlenecks. This may only be true for the realm of network infrastructures (cf. Knieps, 2006a, pp. 59ff.). For instance airport infrastructures, unlike airplanes, involve irreversible costs, because once investments in terminals and runways are made, they cannot be transferred to a different location. In addition, airports usually constitute natural monopolies. Therefore airports are monopolistic bottlenecks.

In European countries railway track infrastructures (unlike railway transport services or railway traffic control) also constitute monopolistic bottlenecks, because track providers, both for long-distance networks and local networks, have a natural monopoly, and building railway tracks involves irreversible costs. However, in the US rail-to-rail competition among railroad companies owning different tracks is a long-run phenomenon (cf. Knieps, 2014). In the telecommunications sector the local loop is the only area where regulation may still be necessary. Monopolistic bottlenecks can be found exclusively in the access networks, while in long-distance networks both active and potential competition exists (cf. Knieps, 2006c).

In the electricity industry sector-specific regulation is necessary in the area of transmission and distribution networks, because these networks are characterised by natural monopoly and irreversible costs. In contrast, generation and supply do not show the characteristics of monopolistic bottlenecks (cf. Knieps, 2006a, pp. 61ff.).

The application of the theory of monopolistic bottlenecks to the network sectors mentioned here can be illustrated by Table 8.2.

Table 8.2 Monopolistic bottlenecks in selected network sectors

	Natural monopoly	Irreversible costs
(a) Air transport		
Supply of air transport services	Yes/no	No
Building and operating air traffic control systems	Yes	No
Building and operating airports	Yes	Yes
(b) Railway		
Supply of railway transport services	Yes/no	No
Building and operating railway traffic control systems	Yes	No
Building and operating railway track infrastructures	Yes	Yes
(c) Electricity		
Generation (production)	No	Yes
Long-distance networks (transmission networks)	Yes	Yes
Regional/local networks (distribution networks)	Yes	Yes
Supply	No	No
(d) Telecommunications		
Terminal equipment	No	No
Telecommunications services (including voice telephone services)	No	No
Satellite/mobile networks	No	No
Long-distance cable-based networks	No	Yes
Local cable-based networks	Yes/no	Yes

8.3 Disaggregated Market Power Regulation

8.3.1 Monopolistic Bottlenecks and the Concept of the Essential Facility

In those network industries which were traditionally vertically integrated (e.g. railroads, electricity, telecommunications) the concept of mandatory third party access became particularly important. The question to be considered is under what criteria the owner of an infrastructure has to open his infrastructure capacities for competitors, and if so, under what conditions. Although the theory of monopolistic bottlenecks holds also for the cases where network infrastructure and the provision of services are vertically separated (e.g. airports), this theory has important roots originating in US Antitrust policy focussing on mandatory third party access. The essential facility doctrine has its roots in the 1912 Terminal Railroad Case,[6] after a merger of railroad terminal facilities which were absolutely

---

[6] U.S. Supreme Court, US v. Terminal Railroad ASS'N of ST. Louis, 224 U.S. 383 (1912), 224 U.S. 383, United States of America, Appt., v. Terminal Railroad Association of ST. Louis et al. No. 386, Decided April 22, 1912.

necessary for all railway companies wanting to pass through or enter St. Louis. Based on the Sherman Act of 1890 the railroad companies owning this integrated terminal were obliged to provide non-discriminatory access to the terminal to all competing railroad companies upon just and reasonable terms.[7]

A facility or infrastructure is termed essential, if it

- is indispensable for reaching customers and/or making it possible for competitors to do business,
- is not otherwise available in the market, and
- objectively cannot be duplicated by economically reasonable means.

This concept is related to the essential facilities doctrine which originated in American antitrust law and is now increasingly also applied in European competition law. This doctrine states that a facility should only be regarded as essential when the following conditions are fulfilled, namely: Entry to the complementary market is effectively not possible without access to this facility; a supplier on the complementary market cannot,[8] with reasonable effort, duplicate the facility; and there are no substitutes (cf. Areeda & Hovenkamp, 1988).

In the context of the disaggregated regulatory approach, the essential facilities doctrine is no longer applied on a case-by-case basis—as was common in antitrust law—but for a class of cases of infrastructures characterised by a natural monopoly in combination with irreversible costs. All areas in a given network sector which have the characteristics of a monopolistic bottleneck are subject to sector-specific market power regulation. The design of non-discriminatory conditions of access to the monopolistic bottleneck facilities must be specified in the context of the disaggregated regulatory approach. The type and extent of the monopolistic bottleneck areas varies considerably between the individual network sectors. It must be shown in detail in which network areas the criteria of a monopolistic bottleneck are in fact fulfilled. In the process, it is important to avoid the danger of misidentifying a monopolistic bottleneck and to apply a concise definition of its boundaries. The application of the monopolistic bottleneck theory has to be seen in a dynamic context. If, for instance due to technological progress, the preconditions for a monopolistic bottleneck no longer exist, the corresponding sector-specific regulation must also be phased out (cf. Knieps, 2011a, p. 21).

8.3.2 Case Study: Newspaper Delivery Service

An interesting example illustrating the danger of misidentifying an essential facility is the so-called "Bronner" case. The crucial question in this case was whether a

---

[7] For a more detailed history of the Terminal Railroad Case in the context of competition among lines in contrast to competition on the track, see Knieps (2014).

[8] Thus it is, for example, not possible to provide a ferry service without access to ports.

home-delivery scheme for newspapers constitutes an essential facility, a question that was examined in detail by the European Court in its Bronner judgment.[9] The newspaper group Mediaprint had developed a nationwide home-delivery system in Austria for its two high-circulation daily newspapers, which allowed early morning delivery of the papers directly to the subscribers. The Bronner firm aspired to have its own daily paper (with lower circulation) included in Mediaprint's home-delivery service. This demand was refused by the European Court, in particular with the argument that (paragraph 44):

> Moreover, it does not appear that there are any technical, legal or even economic obstacles capable of making it impossible, or even unreasonably difficult, for any other publisher of daily newspapers to establish, alone or in cooperation with other publishers, its own nationwide home-delivery scheme and use it to distribute its own daily newspapers.

This argument basically denies that a monopolistic bottleneck facility is involved in establishing nationwide home-delivery systems for daily newspapers. Depending on circulation figures, the geographical distribution of subscribers etc., there are manifold possibilities to create home-delivery systems, even if the option of alternative distribution channels (newspaper kiosks, postal delivery) is not taken into consideration. That economies of scale with regard to delivery exist, due to large circulation, was rightly not accepted as an argument for the existence of an essential facility (paragraph 45):

> It should be emphasised in that respect that, in order to demonstrate that the creation of such a system is not a realistic potential alternative and that access to the existing system is therefore indispensable, it is not enough to argue that it is not economically viable by reason of the small circulation of the daily newspaper or newspapers to be distributed.

The Court's decision also examined the hypothetical case where access to the existing system could be regarded as indispensible. The minimum prerequisite for this would be to prove that it would be unprofitable to create a second home-delivery system for daily newspapers with a comparable circulation. But even this hypothetical case is, from a network economic perspective, not a convincing argument for the existence of an essential facility. In such a case, the economies of scale involved in building an integrated home-delivery system (in relation to the entire market of newspaper subscription) would be so significant that sale via separate home-delivery systems would become unstable. However, as no single newspaper firm can monopolise the scarce inputs needed for building home-delivery systems (staff, vehicles, computer logistics, buildings), due to potential competition, one single efficient home-delivery system would prevail in the market under such conditions. Insisting on a separate home-delivery system would not be incentive-compatible for either of the two newspaper firms, as it would lead to inefficient cost duplication.

---

[9] EuGH, Rs. C-7/97, Urt. v. 26. 11. 1998; see also: Opinion of Advocate General Jacobs of 28/05/98 (published under: http://curia.europa.eu/en/content/juris/c2.htm (viewed 20 May 2014)).

The conclusion to be drawn from this is that even if the building of a specialised delivery system did fulfil the conditions of a natural monopoly, it would not follow that it had the characteristics of a monopolistic bottleneck facility. Lacking active competition is replaced by the disciplinary effects of potential competition.

8.3.3 Limiting Regulation to Monopolistic Bottlenecks

It is necessary to draw a fundamental distinction between the existence of network-specific market power due to monopolistic bottlenecks, and the problem of a possible transfer of this market power to complementary areas. Even if a transfer of market power from a monopolistic bottleneck part to other market segments would be incentive compatible, it does by no means follow that the monopolistic bottleneck part and the other market segments belong to the same relevant market. From the disaggregated regulatory approach it follows that it is necessary to differentiate between network areas with network-specific market power, and network areas characterised by active and potential competition.

Insofar as monopolistic bottlenecks exist in network sectors, they require specific targeted regulation in order to discipline network-specific market power. In particular, symmetric access to monopolistic bottleneck areas for all active and potential suppliers of network services must be guaranteed, so that competition can become fully effective on all complementary markets.

The effect of a total refusal of access to monopolistic bottleneck facilities can also be achieved by providing access only at prohibitively high tariffs. This shows that a regulation of access conditions to monopolistic bottlenecks is necessary. However, the fundamental principle of such a regulatory policy should be to strictly limit regulatory measures to those network areas where market power potential does indeed exist. A regulation of access tariffs to monopolistic bottlenecks must therefore not lead to a simultaneous regulation of tariffs in network areas without market power potential.

Even if, due to the network characteristic, the monopolistic bottleneck areas are complementary to the other network areas, it is not possible to derive from this the necessity for end-to-end regulation, and thus an across-the-board application of the regulatory instruments. The attempt to imitate competition via a "suitable" end-to-end regulation cannot replace a comprehensive deregulation beyond the monopolistic bottlenecks. Only by targeted bottleneck regulation is it possible to quickly identify and institutionally implement the phasing-out potentials of sector-specific regulation.

8.3.4 Anticompetitive Price Structure Regulation

During the era of legally protected monopolies, the objective of eliminating or reducing monopoly rents has led to a multitude of detailed regulatory interventions. In particular, input-based regulatory instruments such as rate of return regulation and mark-up regulation have a long tradition. Input distortions as a side-effect of these regulations (for example the Averch–Johnson effect of excessive capital intensity) are well-know (cf. Braeutigam, 1981). With the beginning of entry liberalisation a paradigm shift towards incentive regulation can be observed. Nevertheless cost-based regulatory rules are also applied by regulatory agencies.

8.3.4.1 Distortion of Competition Through Cost-Based Regulatory Rules

In liberalised network industries the problem of cost recovery has led to different approaches of cost-based regulation; however, all these approaches have to be judged as more or less anticompetitive. A regulatory requirement to impose network access charges in accordance with long-run incremental costs, for instance, would lead to discrimination against the network owner, because nobody would have an incentive to provide network access capacity under these conditions. It can be assumed that the facility would never have been built, if such regulatory requirements had already been under consideration ex ante. Thus there remains the task of also recovering the difference between total costs and incremental costs (that is, the non-attributable costs); the entrepreneurial alternatives of competitors building the facility themselves—and the resultant competitive pressure—already imply an effective upper price ceiling.

A fundamental mistake of fully distributed cost-based regulatory approaches occurs, if regulatory agencies formulate administrative rules specifying the way in which non-attributable costs (or overhead costs) must be allocated. This violates the principle of decision-relevant cost allocation (cf. Sect. 2.2). Thus the so-called 'competition on equal terms rule' mandates that non-attributable costs (overhead costs) must be allocated proportionally to the incremental costs of different services, so that the relative mark-up is identical all over (cf. e.g. Tye, 1993, pp. 46f.). Even if the objective of this allocation rule is equal treatment for all market participants, it is still anticompetitive. In particular, a symmetrical distribution of overhead costs can create incentives for inefficient by-pass of network areas. For instance, if the stand-alone costs of a specialised market entrant are lower than the incremental costs of service provision plus the symmetrically allocated overhead costs, a bargaining solution could be found that specified a lower mark-up, which would, however, still yield a positive cost recovery contribution. In contrast, the competition on equal terms rule leads to incentives for newcomers to provide the necessary services autonomously and thus to inefficient cost duplication.

A second regulatory rule is the so-called 'Efficient Component Pricing Rule' (ECPR) (cf. Baumol, 1983; Baumol & Sidak, 1994) which states that network access charges should not only cover the incremental costs of access, but also the so-called 'opportunity costs of market entry', due to lost revenue of the incumbent firm in the complementary network sectors. Even if before market opening only total cost recovery (without economic profits) was achieved, this rule may create incentives for inefficient bypass. The lower the stand-alone costs of specialised market entry, the greater the likelihood of the entrant foregoing network access as a consequence of this regulatory rule.

The following conclusions can be drawn:

- Pricing that is rigidly tied to long-run incremental costs would excessively limit entrepreneurial flexibility in pricing. This can easily be illustrated by the basic principle of peak load pricing (cf. Sect. 4.1). Because demand for network services varies considerably over time, prices dependent on capacity utilisation can be expected under competition, too. While peak load prices are higher than long-run incremental costs under certain conditions, low-peak costs, conversely, are below long-run incremental costs. Under certain assumptions it is even possible that only short-run marginal costs are compensated in low-peak periods, while all other costs (capacity costs etc.) are covered by peak-load prices.[10] The concept of long-run incremental cost is therefore no suitable pricing parameter for the utilisation of services dependent on capacity utilisation. Instead, entrepreneurs must have the flexibility to either exceed or go below long-run incremental costs.
- A firm that for each of its products only earns the incremental costs cannot recover its product-group specific common costs and its firm-specific common costs (overhead costs)[11] and must inevitably leave the market. The point of reference for competitive prices when there are economics of scale and scope involved thus must be the stand-alone costs. Under competition, the firms themselves are interested in exploiting the signals of marginal costs and long-run incremental costs for their production decisions and determining mark-ups for cost-recovery depending on market conditions (demand elasticities). Governmental regulation with the objective of administratively allocating costs would be detrimental to competition. The conclusion to be drawn from this is that in a competitive context (short-run) marginal costs constitute the lower price limit, while stand-alone costs determine the price ceiling (cf. Baumol & Willig, 1983).

8.3.4.2 Regulation on the Basis of Analytical Cost Models?

Analytical cost models for regulatory purposes were first developed to determine the necessary subsidy for socially desirable universal service provision in local telecommunications networks; then they were utilised to answer the question whether economies of scope could be proven to exist in local telecommunications networks, and finally they were increasingly applied to determine the decision-relevant (incremental) costs of providing network access and interconnections services (cf. e.g. Gabel & Kennet, 1994).

The results of analytical cost models are crucially dependent on the underlying model assumptions, with the concrete design being under debate. In particular,

---

[10] This is the firm peak load case (cf. Steiner, 1957).

[11] It should be noted, however, that the term overhead costs has been applied in traditional literature more broadly including all costs not directly attributable to single products (cf. Clark, 1923).

there are different views on the correct modelling of the network and the technologies employed, as well as on how to model demand. The location of the network nodes, as well as the network hierarchy, the rules of traffic routing etc. can be endogenously chosen within these models.

There are controversies not only regarding the manner of determining and documenting decision-relevant long-run incremental costs, but also regarding who can and should determine these costs. The separation hypothesis states that the costs of efficient service provision constitute an independent parameter by which to measure the actual costs of the incumbent. This separation hypothesis, however, is not compatible with determining decision-relevant costs. The actual decision situation of new entrants and of incumbents has to take the path dependency (expansion versus building from scratch) of the existing network into account (cf. Sect. 2.3.2; Knieps, 2001, pp. 46ff.).

The reference parameter of measuring the costs of efficient service provision on the basis of the costs of a hypothetical market entrant's hypothetical network creates misleading incentives for incumbents as well as market entrants. Potential entrants do not have an incentive to actually enter the market with a more efficient technology, if they can use the established provider's network capacities under the same terms; even less so, if they have to be afraid that in the following period their new technology might already be devalued by still more efficient (hypothetical) technologies, because the incumbent would then have to provide network access under those new terms. The incumbents, on the other hand, would not have any incentives whatsoever to invest further in their networks, because under these conditions they could not expect to recover their cost of capital (cf. Sidak & Spulber, 1998, S. 117).

Network providers will not tolerate being tied to a specific network architecture and specific technologies. After all, network variety and technological variety constitute the motors of dynamic network development. Only active providers can manage the gradual process of further developing their networks in the context of entrepreneurial decisions.

In order to determine the decision-relevant costs of new network subparts, in-house analytical cost models can yield valuable support for the network provider's decision making (cf. Sect. 2.3.4). Analytical cost models, however, cannot reproduce the entrepreneurial decision situations of real network providers. Therefore they are unfit to serve as a basis for determining cost-based tariffs. The reference point of as-if competition must always relate to real entrepreneurial decision situations. External cost models are not able to accomplish this, because they are not geared to reality and thus provide a wrong point of reference (cf. Knieps, 2001, pp. 46ff.).

8.3.5 Price Level Regulation of Access Tariffs

Since the comprehensive opening of networks, price-cap regulation has been the central regulatory instrument. The basic concept of price-cap regulation is

8.3 Disaggregated Market Power Regulation

relatively simple. The assumptions are that there is no perfect regulatory instrument and that regulation can never lead to a perfect correction of market failure. Simplicity and practical applicability are regarded as very important. Regulation should be limited to monopolistic bottlenecks. Even without information on cost and demand conditions, regulation can make things better for customers by making sure that their situation does not become worse. In particular, price level changes of monopolistic bottleneck capacities should only depend on inflation and (expected) change of productivity. On principle, customers should be able to buy an identical amount of different access services in a given basket of services at the same price level (adjusted for inflation and productivity changes) as in the preceding period. *RPI-X* is established as a correction factor, whereby *RPI* constitutes the changes in the consumer price index, and *X* is a percentage to be negotiated between the regulator and the firms, which was in the following interpreted as the percentage of the expected productivity changes in the sector under regulation (cf. Littlechild, 1983, pp. 34–36; Beesley & Littlechild, 1989, pp. 456ff.).

Let $p_{i,t}$ denote the price of the *i*th product in period *t* and $q_{i,t-1}$ the quantity of the *i*th good sold in period $t-1$, then the price-cap constraint is:

$$\sum_{i=1}^{n} p_{i,t} \cdot q_{i,t-1} \leq \sum_{i=1}^{n} p_{i,t-1} \cdot q_{i,t-1} \cdot (1 + RPI - X) \tag{8.1}$$

For this so-called "tariff basket" there exist different variants in regulatory practice; among those, the so-called "average revenue cap" variant proved particularly popular.[12] For this, sales in the regulated area are divided by a homogenous reference figure (for example megawatt hours for electricity networks, number of passengers for airports). Changes in price are permitted, as long as the expected average revenue per output unit in the next year does not exceed the maximum permitted average revenue $\bar{p}$, corrected by the factor $RPI - X$. Thus the following must be valid:

$$\frac{\sum_{i=1}^{n} p_{i,t} \cdot q_{i,t}}{\sum_{i=1}^{n} q_{i,t}} \leq \bar{p} \cdot (1 + RPI - X) \tag{8.2}$$

In contrast to the tariff basket approach, which is based on the observable output quantities of the preceding period, in this case a prognosis regarding the output quantities in period *t* on the part of the regulated firm is required. This increases the potentials for strategic behaviour by the regulated firm.

---

[12] A more detailed and analytical examination of this issue can be found in Bradley and Price (1988) and Beesley and Littlechild (1989, p. 463).

Price-cap regulation is a form of price level regulation which should be limited to monopolistic bottleneck areas. From the perspective of a disaggregated regulatory approach it is a suitable regulatory instrument. Because the prices of the previous period can be expected to be known (at least to the regulatory agency) and *RPI* as well as *X* are exogenously given parameters, detailed information about the cost and demand conditions of the regulated firm is not necessary. Because price-cap regulation lets the regulated firm keep efficiency gains in the form of cost savings, incentive distortions of the sort caused by input regulation do not occur.

Price-cap regulation is an innovative regulatory instrument that can be utilised for disciplining remaining network-specific market power in monopolistic bottleneck areas. However, this does not mean that price-cap regulation is a perfect regulatory instrument. In particular, the danger arises that price-cap regulation is applied in those network areas which are competitive. Therefore a phasing out of sector-specific regulation should be implemented as swiftly and comprehensively as possible, as soon as the market power in need of being disciplined does no longer exist and competition can function.

Price-cap regulation can only create incentives for increasing efficiency because it allows the regulated firm to keep at least part of the profits of efficiency gains, so that even with price-cap regulation economic profits in monopolistic bottlenecks do not completely disappear. This is something price-cap regulation has in common with other forms of incentive-based regulation (cf. e.g. Vogelsang, 2002).

The following basic principles for price level regulation can be summarised:

- Regulation should be strictly limited to verified market power in monopolistic bottleneck areas. Over the course of time the continuing existence of these areas should be regularly checked und regulation should be stopped at once when a bottleneck no longer exists.
- The point of reference in the sense of as-if competition, where accusing the owner of a monopolistic bottleneck facility of abusing market power is not justified, should be the recovery of the total costs of the monopolistic bottleneck.
- Regulatory authorities should not obligate firms to use specific pricing rules for access tariff systems. This would obstruct the entrepreneurial search for innovative tariff systems (cf. Sect. 4.1.2).
- Price-cap regulation of monopolistic bottleneck areas together with accounting separation are sufficient to discipline the remaining market power und guarantee non-discriminatory access to monopolistic bottleneck facilities. By limiting the scope of regulatory rules to the level of output prices the regulatory agency's information requirements are kept to a minimum. This not only reduces regulatory effort, but also creates entrepreneurial incentives to seek out cost savings and innovative price structures. The decisive advantage of price-cap regulation over the individual rate approval procedure is the fact that the former does not impede the entrepreneurial quest for innovative price structures.

8.3.6 Implementation of Price-Cap Regulation

The decisive parameter in price-cap regulation is the X-factor, which is determined by the regulatory agency and which prescribes how much the level of output prices (corrected for inflation) can be raised or must be lowered. The basic idea of price-cap regulation is to determine the X-factor—which is exogenously prescribed for the regulated firms—as the estimation of the growth rate of total factor productivity of the regulated network area.

In Great Britain different X-factors were calculated for the individual regulated sectors. The highest value for X was determined for the telecommunications sector, due to strong technological progress. The lowest X-factor emerged in the water industry, where significant increases of price levels were permitted. Over the course of time, X-factors were increased significantly for both airports and gas networks (c.f. Rees & Vickers, 1995, p. 376).

Of late, in particular in the electricity sector, firm-specific X-factors have been increasingly applied, with the variable of productive efficiency for each individual firm being determined by benchmarking. The measurability of productive efficiency in individual firms as compared to other firms was already examined in theory in a groundbreaking study by Farrell (1957). A distinction is made between technical efficiency (producing maximum output with given input) and price efficiency (cost minimising adjustment to input price changes). In the following, a multitude of different econometric procedures and approaches for linear optimisation were developed.[13]

A fundamental distinction has to be made between the application of firm-specific X-factors for intra-firm purposes and the application of firm-specific X-factors for price-cap regulation.[14] Firm-specific X-factors run counter to the basic concept of price-cap regulation, because incentive regulation is here partially or completely replaced by an ex ante firm-specific cost control. In addition, firm-specific X-factors determined on the basis of a benchmark are problematic when used for regulatory purposes (cf. e.g. Shuttleworth, 2005; Ajodhia, Petrov, & Scarsi, 2003; Turvey, 2006):

- The results of a benchmark are crucially dependent on the choice of model, model specifications and on data quality. Applying a credible regulatory instrument, on the other hand, requires robust estimations of (in-)efficiencies which are relatively independent of the choice of model specifications.
- A further difficulty with regard to benchmarking is how to separate unobservable factors of heterogeneity between different networks and the resultant cost differences from real inefficiencies. Thus the costs of, for example, electricity

---

[13] For an overview of the benchmarking literature cf. e.g. Riechmann and Rodgarkia-Dara (2006), Murillo-Zamorano (2004).

[14] Recently, determining the "Enterprise Total Factor Productivity" has gained increasing importance as a new approach in corporate management; cf. e.g. Lev and Daum (2003, p. 9).

distribution networks depend on a large number of exogenous factors (e.g. topography, distribution of demand over space, etc.) which can vary significantly between different distribution networks. Consequently, the providers of electricity distribution networks work under different conditions, so that cost differences are to be expected, and must not be equated with cost inefficiencies.

- Finally, there is a considerable range for periodising decision-relevant cost of capital, depending on the firm-specific expectations regarding the future development of input and output prices, technological progress, etc. This can in particular lead to different economic depreciation procedures in the period under examination (cf. Sect. 2.1.2). It would be misleading to regard these differences as inefficiencies; it would be equally misleading to mandate the use of a unified depreciation method.

8.4 Questions

8-1: Network-Specific Market Power
What are the conditions for a network area to not be a monopolistic bottleneck facility?

8-2: Potential for Competition on Transport Markets
Explain the preconditions for competition on the markets for transport services.

8-3: Price Level Regulation of Access Tariffs
Explain the basic principles of price-cap regulation.

References

Ajodhia, V., Petrov, K., & Scarsi, G. C. (2003). Benchmarking and its applications. *Zeitschrift für Energiewirtschaft, 27*(4), 261–274.

Areeda, P., & Hovenkamp, H. (1988). "Essential facility" doctrine? Applications. *Antitrust Law, 202.3*(Suppl. 1988), 675–701.

Baumol, W. J. (1982). Contestable markets: An uprising in the theory of industry structure. *American Economic Review, 72*, 1–15.

Baumol, W. J. (1983). Some subtle issues in railroad regulation. *International Journal of Transport Economics, 10*(1–2), 341–355.

Baumol, W. J., & Sidak, G. (1994). *Toward competition in local telephony*. Cambridge, MA: AEI Studies; MIT Press.

Baumol, W. J., & Willig, R. D. (1983). Pricing issues in the deregulation of railroad rates. In J. Finsinger (Ed.), *Economic analysis of regulated markets* (pp. 11–47). London: McMillan.

Beesley, M. E., & Littlechild, S. C. (1989). The regulation of privatised monopolies in the United Kingdom. *Rand Journal of Economics, 20*, 454–472.

Blankart, C. B., & Knieps, G. (1995). Market-oriented open network provision. *Information Economics and Policy, 7*, 283–296.

Bradley, I., & Price, C. (1988). The economic regulation of private industries by price constraints. *Journal of Industrial Economics, XXXVII*(1), 99–106.

References

Braeutigam, R. R. (1981). Regulation of multiproduct enterprises by rate of return, mark-up, and operation ratio. *Research in Law and Economics, 3*, 15–38.

Clark, J. M. (1923). *Studies in the economics of overhead costs*. Chicago: University of Chicago.

Demsetz, H. (1968). Why regulate utilities? *Journal of Law and Economics, 11*, 55–65.

European Commission (2011). *Connecting Europe: The new EU core transport network*. MEMO/ 11/706 19/10/2011, Europa.eu, Press releases database, http://europa.eu/rapid/press-release_ MEMO-11-706_en.htm. May 20, 2014.

Farrell, M. J. (1957). The measurement of productivity efficiency. *Journal of the Royal Statistical Society, Series A (General), 120*(III), 253–281.

Gabel, D., & Kennet, D. M. (1994). Economies of scope in the local telephone exchange market. *Journal of Regulatory Economics, 6*, 381–398.

Joskow, P. L., & Klevorick, A. K. (1979). A framework for analyzing predatory pricing policy. *Yale Law Journal, 89*, 213–270.

Knieps, G. (1997). Phasing out sector-specific regulation in competitive telecommunications. *Kyklos, 50*(3), 325–339.

Knieps, G. (2001). Costing and pricing in liberalised telecommunications markets. In J. G. Sidak, C. Engel, & G. Knieps (Eds.), *Competition and regulation in telecommunications – examining Germany and America* (pp. 41–49). Boston: Kluwer.

Knieps, G. (2006a). Sector-specific market power regulation versus general competition law: Criteria for judging competitive versus regulated markets. In F. P. Sioshansi & W. Pfaffenberger (Eds.), *Electricity market reform: An international perspective* (pp. 49–74). Amsterdam: Elsevier.

Knieps, G. (2006b). Privatisation of network industries in Germany: A disaggregated approach. In M. Köthenbürger, H.-W. Sinn, & J. Whalley (Eds.), *Privatisation experiences in the European Union* (pp. 199–224). Cambridge, MA: MIT Press.

Knieps, G. (2006c). The different role of mandatory access in German regulation of railroads and telecommunications. *Journal of Competition Law and Economics, 2*(1), 149–158.

Knieps, G. (2006d). *Delimiting regulatory needs*. In OECD/ECMT Round Table 129, transport services: The limits of (De)regulation, Paris, pp. 7–31.

Knieps, G. (2011a). The three criteria test, the essential facilities doctrine and the theory of monopolistic bottlenecks. *Intereconomics, 46*(1), 17–21.

Knieps, G. (2011b). Network neutrality and the evolution of the internet. *International Journal of Management and Network Economics, 2*(1), 24–38.

Knieps, G. (2014). Competition and third party access in railroads, forthcoming: In M. Finger, T. Holvad & P. Messulam (Eds.), *Rail economics, policy and regulation in Europe*. Cheltenham: Edward Elgar.

Knieps, G., & Zenhäusern, P. (2010). Phasing out sector-specific regulation in European telecommunications. *Journal of Competition Law and Economics, 6*(4), 995–1006.

Knieps, G., & Zenhäusern, P. (2014). Regulatory fallacies in global telecommunications: The case of international mobile roaming. *International Economics and Economic Policy, 11*(1), 63–79.

Landes, W. M., & Posner, R. A. (1981). Market power in antitrust cases. *Harvard Law Review, 94*, 937–997.

Lev, B., & Daum, J. H. (2003). Intangible assets: Neue Ansätze für Unternehmenssteuerung und Berichtwesen. In P. Horváth & R. Gleich (Eds.), *Neugestaltung der Unternehmensplanung* (pp. 33–49). Stuttgart: Schäffer/Poeschel-Verlag.

Littlechild, S. C. (1983). *Regulation of British telecommunications' profitability*. London: Department of Industry.

Murillo-Zamorano, L. R. (2004). Economic efficiency and frontier techniques. *Journal of Economic Surveys, 18*(1), 33–77.

Panzar, J. C., & Willig, R. D. (1977). Free entry and the sustainability of natural monopoly. *Bell Journal of Economics, 8*, 1–22.

Rees, R., & Vickers, J. (1995). RPI–X price-cap regulation. In M. Bishop, J. Kay, & C. Mayer (Eds.), *The regulatory challenge* (pp. 358–385). Oxford: Oxford University Press.

Riechmann, C., & Rodgarkia-Dara, A. (2006). Regulatorisches Benchmarking – Konzeption und praktische Interpretation. *Zeitschrift für Energiewirtschaft, 30*(3), 205–219.

Schmalensee, R. (1989). Inter-industry studies of structure and performance. In R. Schmalensee & R. Willig (Eds.), *Handbook of industrial organisation* (pp. 951–1009). Amsterdam: North-Holland.

Shuttleworth, G. (2005). Benchmarking of electricity networks: Practical problems with its use for regulation. *Utilities Policy, 13*, 310–317.

Sidak, J. G., & Spulber, D. F. (1998). Deregulation and managed competition in network industries. *Yale Journal on Regulation, 15*(1), 117–147.

Steiner, P. O. (1957). Peak loads and efficient pricing. *Quarterly Journal of Economics, 71*, 585–610.

Tirole, J. (1989). *The theory of industrial organisation*. Cambridge: MIT Press (2nd printing).

Turvey, R. (2006). On network efficiency comparisons: Electricity distribution. *Utilities Policy, 14*, 103–113.

Tye, W. B. (1993). Pricing market access for regulated firms. *Logistics and Transportation Review, 29*(1), 39–67.

Vogelsang, I. (2002). Incentive regulation and competition in public utility markets: A 20-year perspective. *Journal of Regulatory Economics, 22*(1), 5–27.

Walras, L. (1874/1877). *Eléments d'économie politique pure ou Théorie de la richesse sociale*. L. Corbaz, Lausanne.

Weizsäcker, C. C. von (1984). The costs of substitution. *Econometrica, 52*(5), 1085–1116.

The Positive Theory of Regulation

9.1 Normative Versus Positive Theory of Regulation

The normative theory of regulation defines the criteria for determining which network areas should be regulated (regulatory basis), which instruments should be used to do so (regulatory instruments) and how these instruments should be applied. The central question is how network industries should be regulated. In contrast, the positive theory of regulation studies the emergence, the transformation, and the abolishment, as well as the institutional implementation of sector-specific regulation. Thus the central question is how network industries actually are regulated. In examining this question, the influence exercised by firms, consumer interests, and the bureaucratic self-interest of the regulatory agency must be taken into account in order to explain the behaviour of regulators.[1] The various interest groups of consumers and producers as well as other related interest groups (pursuing e.g. environmental, health, climate, landscape conservation interests) compete with each other in their quest for political influence. The influence exerted by the interest groups involved is fundamentally dependent on the institutional framework.

The starting point of the normative theory of regulation is the concept of market failure or market imperfections in need of being corrected by regulatory interventions. Technical regulations pursuing safety and coordination issues are required to enable the functioning of the market processes. Universal service objectives are implemented by regulation (e.g. a universal service fund), if the provision of a socially desired network service at an "affordable" price cannot be guaranteed via the market process. Since network-specific market power can exist even after the complete opening of the network industries, ex ante market power regulation is required, complementary to the general competition law. For a long

---

[1] An overview of positive theories of regulation can be found in Joskow and Noll (1981), Owen and Braeutigam (1978), Spulber (1989), Vogelsang (1988).

time the public interest theory of regulation assumed that the formulation and the implementation of laws served public interest exclusively, without any interference by the regulatory agency's self-interest. The dissatisfactory results that could be observed in regulatory practice led initially to a reformulation of public interest theory and to the assumption that, although regulatory agencies had been created to protect public interest, because of mismanagement they did not always achieve this objective. However, this re-formulation neglected the influence that interest groups exercise upon regulatory agencies, so that even agencies that work efficiently may pursue other objectives. This finally led to a fundamental critique of the public interest theory of regulation and to the development of the so-called positive theory of regulation with its focus on the role of interest groups in the regulatory process (cf. e.g. Posner, 1974, pp. 336ff.).

9.2 The Positive Theory of the Behaviour of Regulatory Agencies

9.2.1 The Cornerstones of the Regulatory Process

Regulatory legislation typically consists of two components, the legal framework of regulation and the legal regulatory mandate. Within the legal framework of regulation, regulatory agencies possess a certain discretionary freedom of action as regards the implementation of regulation and the choice and application of the regulatory instruments. This leads to the problem of the legal regulatory mandate which restricts the regulatory agency's freedom of action. In liberalised network sectors there is a need for a legal framework of regulation guaranteeing comprehensive market opening including adequate technical regulations, the provision of universal service compatible with market entry, as well as rules for disciplining network-specific market power. Depending on the specific legal objectives, appropriate regulatory mandates have to be derived which impose a binding order on the relation between legislator and regulatory agency. Within the framework of the legally specified regulatory mandate, regulation is implemented by the application of the regulatory instruments (cf. Fig. 9.1).

9.2.2 The Legal Framework of Regulation

The legal framework of regulation is established in the political process, in the form of laws and regulations. Sector-specific regulation is implemented by agencies established by the legislator. Posner (1974, pp. 339f.) already pointed out the necessity for a division of labour between the legislator and the regulatory agencies implementing the law.

Legislative processes can relate to the introduction, the further development, or the abolishment of specific regulatory rules concerning the extent of regulation and the regulatory instruments to be applied. During the last decades the change from

9.2 The Positive Theory of the Behaviour of Regulatory Agencies

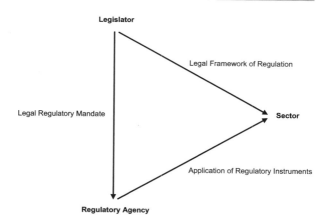

Fig. 9.1 The regulatory triangle

globally entry protected network monopolies to partial entry deregulation followed by global entry deregulation has taken place (cf. Köthenbürger et al. 2006). This process of gradual market opening has been time-consuming and the transformation from global end-to-end regulation of legally protected network monopolies to the full implementation of the disaggregated regulatory framework is still ongoing. Legislation and regulation in opened network sectors apply to technical regulation, universal service regulation and market power regulation. In this context, the regulatory agencies are of particular importance, as they are the institutions executing and implementing the legal framework. The decisive influence of different interest groups has to be taken into account in order to understand the past regulatory reforms and to tackle future challenges.

By transferring regulatory competence from the legislator to a regulatory agency, the future freedom of action of the regulatory agency is simultaneously determined (Spulber & Besanko, 1992, pp. 127ff.). This is a regulatory mandate from the legislator (principal) to the regulatory agency (agent) where the extent of the regulatory agency's freedom of action can be relatively small or large.

If the legal framework does not provide for a regulatory agency, regulatory tasks are either not performed at all, or they are performed by other institutions (competition authorities, courts). This approach, i.e. completely dispensing with sector-specific agencies, is called light-handed regulation. It was practiced in New Zealand for a substantial period of time. In Germany, too, the energy and railway laws were only adapted to introduce sector-specific market power regulation of network access in 2005 (cf. Knieps, 2005b, p. 24, 2006a, p. 204).

The opposite of light-handed regulation is heavy-handed regulation. Here the legislator allows the regulatory agency considerable leeway, both with regard to determining the regulatory basis and in choosing the regulatory instruments. European telecommunications regulation may serve as an illustrative example of heavy-handed regulation (cf. e.g. Knieps, 2005a): This comprehensive freedom of action on the part of the regulatory agencies led to a situation where even competitive network areas—such as long-distance telecommunications networks—were subject to regulation. The choice of certain regulatory instruments, such as, for instance, detailed cost-based price control in combination with price-cap regulation, also resulted in over-regulation in the telecommunications sector. It is only recently that the process of phasing out sector-specific regulation has gained momentum (cf. Knieps & Zenhäusern, 2010).

From the perspective of normative regulatory theory, neither light-handed nor heavy-handed regulation can be recommended (cf. Knieps, 2006b). Therefore it is necessary that the legislator, via a regulatory mandate, defines the regulatory agency's scope of competence in such a way that systematic over- and under-regulation are both avoided. On the one hand, this restricts the regulatory agency's freedom of action and thus prevents unnecessary regulatory interventions; necessary regulatory interventions, on the other hand, are legally prescribed. All regulatory functions (technical regulation, universal service regulation, market power regulation) require a regulatory mandate from the legislator to the regulatory agency. After the necessary technical regulations are installed in the relevant laws, the preconditions for the functioning of market processes are given. However, if adequate technical regulations are missing, market power regulation may be applied as an inadequate substitute for technical regulations (cf. Knieps & Zenhäusern, 2014). In the following, the focus will be on the issue of market power regulation.

9.2.3 The Regulatory Agency's Discretionary Freedom of Action

Regulatory agencies are competent to make decisions within the framework of their legal mandate. It is assumed that regulatory agencies cannot become active beyond the limits of this legal mandate. Therefore the legal mandate constitutes a restriction of the regulatory agencies' behaviour. Within this scope of activity, regulatory agencies possess discretionary freedom of action the extent of which varies considerably, depending on the specific design of the legal framework of regulation. A certain freedom of action is prima facie desirable, and thus promoted by the legislator, because it gives the regulatory agency the necessary flexibility to deal with the specific requirements of implementing sector-specific regulation; in particular, the application of regulatory rules requires specific knowledge of the relevant sector. The extreme case that regulatory laws leave no discretionary freedom of action at all to the regulatory agency can therefore be excluded. The opposite extreme, i.e. that regulatory agencies act completely independently of the legal regulatory framework, is equally unlikely.

Thus it can be assumed that the legal framework of regulation will restrict, but not completely abolish, the discretionary freedom of action of regulatory agencies (cf. Blankart & Knieps, 1989; Spulber & Besanko, 1992; Weingast & Moran, 1983). Usually the legal rules are formulated in such a way that the regulatory agencies can choose from a smaller or larger variety of regulatory alternatives that are compatible with the law. Thus the regulatory agencies are able to direct the regulatory process within the legal framework. This approach is based on the view that the transaction costs of responding to changes that are relevant to the regulation of a certain sector are considerably lower if the response is organised by the decisions of a specialised agency rather than by legal amendments instituted by parliamentary institutions. According to Posner (1974), the sheer size of a parliamentary assembly makes it necessary to delegate periodically returning functions requiring special expertise to specialised institutions. The possibility of parliamentary amendments of legislation concerning network sectors is of course not excluded. Because of the higher cost involved, however, they are less frequent than decisions made by the regulatory agencies.

9.2.4 The Influence of Interest Groups

9.2.4.1 Ad-hoc Hypotheses

Several ad hoc hypotheses on how successfully a specific interest group will be able to push its agenda vis-à-vis a regulatory agency are known in the literature.

- The Consumer Interest Groups Hypothesis
  The consumer interest groups hypothesis states that regulatory agencies, mandated by the legislator, focus on consumer surplus exclusively and neglect the firms' profits in their decision making. It is obvious that this hypothesis has its roots in the public interest theory of regulation holding that government regulations are costless and efficient and that regulatory agencies have the proper incentives to pursue welfare-maximising goals without setting their own agenda. This hypothesis, however, is unsuitable for predicting regulatory agencies' behaviour (cf. Posner, 1974, pp. 336ff.). In particular, it fails to examine the interaction of the contradictory interests of different interest groups. Conflicts of interests may occur between consumer groups and producers, but may also arise between different consumer groups.[2]
- The Capture Hypothesis
  The so-called capture hypothesis is a specific explanation of regulatory interventions that are aimed at creating and redistributing rents. According to a well-known version of this hypothesis, regulatory agencies are at first created in the public interest (e.g. in order to serve consumers interest); subsequently, however, they become the tools of the industry they were meant to regulate (cf. e.g. Bernstein, 1955; Owen & Braeutigam, 1978, p. 11).

  Thus this relatively simple theory rests on the assumption that over time regulatory agencies come to be dominated by the regulated industry; the original objectives of the regulatory program are for the most part abandoned, as, for example with regard to profit regulation, the profitability permitted increases

---

[2] The increasing debate on the role of consumer empowerment programs in order to enhance the capacity of consumer representatives to become active during the regulatory proceedings in the US (cf. Schwarcz, 2012) indicates the conflicts of interest of different interest groups involved.

steadily until it approaches monopoly profits. In the end, the regulatory agency's objective function is merely a maximisation of industry profits. However, the capture hypothesis is unsatisfactory, because it lacks a theoretical foundation In particular, it does not constitute a theory of the systematic exercise of influence by different interest groups, including the consumer groups (cf. Posner, 1974, pp. 341ff.).

9.2.4.2 Competition Between Interest Groups

The influence of interest groups on the regulatory process has been analysed by Stigler (1971), Peltzman (1976) and Becker (1983) in the context of the positive theory of regulation. The starting point is the assumption that regulation always leads to a redistribution of rents, so that regulatory measures always result in creating winners and losers. The potential of the redistribution of rents between different interest groups defines the supply side of the regulatory process.

The demand for regulation is created by the influence of interest groups on the regulatory agency. The participants of the regulatory process have an interest in influencing the regulator's decisions in their favour. The competition for rents is frequently linked to the deployment of resources for lobbying activities in order to increase the extent of one's influence.

A simple model approach is sufficient to illustrate the interplay of supply and demand (cf. Peltzman, 1976, pp. 217ff.; Spulber, 1989, pp. 94ff.). The approach is based on a scenario with two interest groups, consumers and producers. Let $\Omega(p)$ denote consumer surplus and $\pi(p)$ industry profit. A special scenario is assumed where the only objective of regulation is redistribution and the regulator has the competency to impose, via barriers to entry and price regulation, a price $p^c \leq p^a \leq p^m$.[3] This scenario results in a potential for rent redistribution. It can also be interpreted as a regulatory production possibilities set, the efficient margin of which is characterised by the production possibilities frontier T with $T(\pi(p), \Omega(p)) = 0$ (cf. Fig. 9.2).[4] Regulated profit $\pi(p^a)$ ranges between 0 and monopoly profit π^m. Regulated consumer surplus $\Omega(p^a)$ ranges between $\Omega(p^m)$ and $\Omega(p^c)$.

From the total differential of $T(\pi(p), \Omega(p))$

$$\frac{\partial T}{\partial \pi(p)} \cdot d\pi(p) + \frac{\partial T}{\partial \Omega(p)} \cdot d\Omega(p) = 0 \qquad (9.1)$$

it follows that the marginal rate of regulatory rent distribution is:

---

[3] The crucial contribution of the Stigler/Peltzman model is the explicit identification of the role played by interest groups in the regulatory process. The institutional process itself, however, is not examined in detail; in particular, no distinction is made between legislative institutions and regulatory agencies (cf. Weingast & Moran, 1983, p. 768).

[4] On the concept of the production possibilities frontier, cf. e.g. Varian (2010, pp. 629f.).

9.2 The Positive Theory of the Behaviour of Regulatory Agencies

Fig. 9.2 The Stigler/Peltzman model (based on Spulber, 1989, p. 95, Fig. 2.3.1)

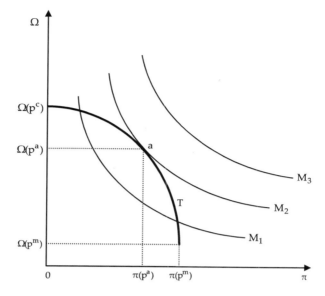

$$MRT = \frac{d\Omega(p)}{d\pi(p)} = -\frac{\frac{\partial T}{\partial \pi(p)}}{\frac{\partial T}{\partial \Omega(p)}} \quad (9.2)$$

In addition, a function M of the influence exercised by the interest groups on the regulator is assumed. The extent of influence exercised is dependent on the intensity of the interests.[5]

The regulator's objective function M can be characterised as follows:

$$M = M(\pi(p), \Omega(p)) \quad (9.3)$$

$$\frac{\partial M}{\partial \pi(p)} > 0 \text{ and } \frac{\partial M}{\partial \Omega(p)} > 0$$

The support of the consumer group increases if regulation enables an increase of consumer surplus. The support of the producer group increases if regulation enables an increase of profits. At the same time, it is assumed that marginal returns of rent redistribution to a particular group decrease.

This means that: $\frac{\partial^2 M}{\partial \pi(p) \partial \pi(p)} < 0$ and $\frac{\partial^2 M}{\partial \Omega(p) \partial \Omega(p)} < 0$

From this it follows that the iso-influence functions are convex. The marginal rate of substitution for a regulation favouring one interest group and disadvantaging the other interest group is:

---

[5] The extent of influence exercised is determined by "dollar votes", i.e. the acceptance of a suggested change depends primarily on the intensity of the influence exerted and not on the number of participants.

$$MRS = -\frac{\frac{\partial M}{\partial \pi(p)}}{\frac{\partial M}{\partial \Omega(p)}} \quad (9.4)$$

The regulator's optimisation problem is given by a maximisation of influence function M of the consumer and producer groups, under the auxiliary condition of the available rent redistribution potential (production possibilities set):

$$\max_p M(\pi(p), \Omega(p)), \quad (9.5)$$

Such that: $T(\pi(p), \Omega(p)) = 0$

The Lagrange function is:

$$L = M(\pi(p), \Omega(p)) + \lambda(T(\pi(p), \Omega(p)) - 0) \quad (9.6)$$

and thus:

$$\frac{\partial L}{\partial \pi(p)} = \frac{\partial M}{\partial \pi(p)} + \lambda \frac{\partial T}{\partial \pi(p)} = 0 \quad (9.7)$$

and:

$$\frac{\partial L}{\partial \Omega(p)} = \frac{\partial M}{\partial \Omega(p)} + \lambda \frac{\partial T}{\partial \Omega(p)} = 0 \quad (9.8)$$

The result is a regulated price p^a such that:

$$MRS = -\frac{\frac{\partial M}{\partial \pi(p)}}{\frac{\partial M}{\partial \Omega(p)}} = -\frac{\frac{\partial T}{\partial \pi(p)}}{\frac{\partial T}{\partial \Omega(p)}} = MRT \quad (9.9)$$

The interests of producers and consumers are balanced in such a way that the marginal rate of regulatory substitution between consumer and producer interests is equal to the marginal rate of transformation of the rent redistribution potential.

The explanatory value of the positive theory is crucially dependent on the information available on influence function M and on the effects of changes in rent redistribution potential caused, for instance, by technological progress (cf. Peltzman, 1976, p. 233). Statements of a general nature on the role of interest groups are scarce. The demand for government intervention, be it regulatory or deregulatory, generally causes costs for the co-ordination of interests. Thus it is not sufficient for a group to know its interests and objectives in order to be successful. It is also necessary for the group to organise, so that it can achieve its objectives in competition against other interest groups. A central insight of the theoretical analyses is that an interest group's actual influence is not determined by its total deployment of resources, but that it instead emerges in the simultaneous interaction of all interest groups involved in the process.

9.2 The Positive Theory of the Behaviour of Regulatory Agencies

It is difficult to formulate valid general statements on the role of group size in the regulatory process. Assuming that organisation costs increase progressively with group size, it appears plausible that small groups can push their agenda more successfully than large groups (Stigler, 1971, 1974). However, this concept of a small group being favoured at the expense of a large group becomes unstable if the assumption of progressively increasing organisation costs is abandoned. The crucial contribution of the positive theory of regulation is its consistent elaboration of the important role played by the competition between interest groups in the regulatory process. Von Weizsäcker (1982, p. 335) already points out the necessity of examining the more general question of different coalitions of interest and their varying abilities to assert themselves in the context of a balance of coalitions. It is thus not necessarily the relative size of interest groups, but rather their ability to form stable coalitions that leads to the successful enforcement of a specific regulatory policy. It is difficult to derive consistent statements on the capabilities of interest groups to exert influence in a specific regulatory context, even by means of game theoretical model analyses (cf. e.g. Becker, 1983); case studies prove to be more suitable.

Other factors besides the influence of interest groups may have an impact on the behaviour of regulatory agencies. According to the theory of bureaucratic behaviour, administrations are above all interested in the maximisation of their budgets (Niskanen, 1971). Thus a regulatory agency may be interested in an expansion of its responsibility and an increase of regulatory complexity, because this leads to a broadening of its field of activity, which in turns justifies the employment of additional staff. Just like managers in a business that maximises profit cannot be completely controlled by the owners, it is possible that for a regulatory agency's objectives, other components (such as size, administrative complexity) besides the influence of interest groups become relevant (cf. Owen & Braeutigam, 1978, pp. 15f.).

In conclusion it should be noted that influence on a regulatory agency is not exerted by a democratic voting process. Rather, the acceptance of a reform proposal depends on how much influence is exerted by the individual participants in the regulatory process. A fundamental differentiation must be made between consumer groups and producer groups. Even within consumer groups, interests may differ; for instance, the interests of major customers may be different from those of small customers. The same is true for the producers, for instance, the interests of incumbents may be different from those of (potential) entrants (cf. Blankart & Knieps, 1989). The extent to which individual interest groups are able to assert their agenda vis-à-vis the regulatory agency depends on the specific conditions of the regulatory process, so that it is very difficult to make general statements about this. It is, however, highly unlikely that the process automatically results in a welfare-maximising scenario.

9.2.5 The Disaggregated Regulatory Mandate

The disaggregated regulatory mandate must be compatible with the legal framework of regulation. It is crucial that both components are defined on the basis of a disaggregated regulatory approach. Reform efforts must start with the design of the legal regulatory mandate, so that regulatory agencies have economically efficient incentives and double regulation, misregulation, over- and under-regulation are avoided. As regards the remaining scope of activity, however, the competency of the relevant regulatory agency should hold. Remaining conflicts among parties in dispute are under the competency of the relevant courts.[6]

9.2.5.1 Basic Elements

The legal framework of regulation and the legal regulatory mandate have to be compatible. Within the legal framework of regulation basic principles such as free market entry, universal service objectives, market power regulation as well as technical regulations are laid down. The recipient is the sector that needs to be regulated. This must be clearly distinguished from the regulatory mandate, the recipient of which is the regulatory agency. Here the competencies of the regulatory agency for implementing regulation are determined, for instance the restrictions regarding the combination of regulatory instruments or the length of price-cap periods. In the following, the basic elements of a regulatory mandate based on disaggregated regulatory economics (disaggregated regulatory mandate) will be presented (cf. Knieps, 2005a, pp. 81ff.).

- Restricting regulation to areas with network-specific market power
  Global regulation that also applies to competitive areas is incompatible with the principle of disaggregated regulation. But temporary or complete suspension of regulation in areas with network-specific market power is equally unjustifiable.
- Phasing out regulation when network-specific market power no longer exists
  As soon as, for instance because of technological progress, network-specific market power disappears in a specific network area, regulation of this area must be terminated also.
- Incentive regulation in monopolistic bottleneck areas
  Non-discriminatory access to monopolistic bottleneck facilities must be guaranteed. Excessive price levels in monopolistic bottleneck areas should be reduced by means of a targeted application of price-cap regulation. At the same time it must be ensured that regulation does not endanger recovery of total cost, including decision-relevant cost of capital.

---

[6] It is interesting to note that landmark decisions of courts (e.g. the European Court of Justice as well as National High Courts) also played an important role in getting the process of regulatory reform running (cf. Knieps, 2014).

9.2.5.2 Strengthening the Regulatory Agency's Self-Commitment Capability

The disaggregated regulatory mandate simultaneously constitutes a binding constraint of the regulatory agency's freedom of action and thus reduces the possibility of opportunistic behaviour. In particular, regulatory agencies are also obliged to enable the compensation of ex ante risks of investment projects, including the risk of failure. Without such an obligation on the level of the legally specified disaggregated regulatory mandate, potential investors will expect regulatory opportunism, an expectation that regulatory agencies will not be able to counter because of their inability to plausibly commit themselves to non-opportunistic behaviour with regard to the regulated activities. This problem is caused by the sequential nature of investment decisions (ex ante) versus the regulation of access tariffs (ex post). As regulatory agencies cannot be assumed to be welfare-maximising, they have relatively strong incentives to only apply price regulation to successful projects, while the ex ante risk of failed projects will not be compensated (cf. Newbery, 2000, pp. 34–36). As investors have no incentive to make investments under these conditions, this inability to plausibly commit on the part of the regulatory agency is used as justification for demanding a temporary suspension of regulation, so-called access holidays (cf. Gans & King, 2003). However, even a temporary abolishment of the regulation of network-specific market power is to be avoided. The disaggregated regulatory mandate is sufficient to plausibly commit regulatory agencies and thus create the necessary investment incentives.

9.3 Questions

9-1: Normative Versus Positive Theory of Regulation
What are the differences between the positive and the normative theory of regulation?

9-2: The Cornerstones of the Regulatory Process
Explain the cornerstones of the regulatory process.

9-3: Interest Groups and Regulatory Agencies
Explain the two contrary ad hoc hypotheses on the assertiveness of interest groups in the regulatory process.

9-4: Disaggregated Regulatory Mandate
Explain the objective and the basic elements of the disaggregated regulatory mandate.

References

Becker, G. S. (1983). A theory of competition among pressure groups of political influence. *Quarterly Journal of Economics, 98*(3), 371–400.

Bernstein, M. H. (1955). *Regulation business by independent commissions*. Princeton: Princeton University Press.

Blankart, C. B., & Knieps, G. (1989). What can we learn from comparative institutional analysis? The case of telecommunications. *Kyklos, 42*(4), 579–598.

Gans, J., & King, S. (2003). Access holidays for network infrastructure investment. *Agenda, 10*(2), 163–178.

Joskow, P. L., & Noll, R. A. (1981). Regulation in theory and practice: An overview. In A. Fromm (Ed.), *Studies in public regulations* (pp. 1–65). Cambridge, MA: MIT Press.

Knieps, G. (2005a). Telecommunications markets in the stranglehold of EU regulation: On the need for a disaggregated regulatory contract. *Journal of Network Industries, 6*(2), 75–93.

Knieps, G. (2005b). Railway (DE-)regulation in Germany. *CESifo DICE Report, 3*(4), 21–25.

Knieps, G. (2006a). Privatisation of network industries in Germany: A disaggregated approach. In M. Köthenbürger, H.-W. Sinn, & J. Whalley (Eds.), *Privatization experiences in the European Union* (pp. 199–224). Cambridge, MA: MIT Press.

Knieps, G. (2006b). Sector-specific market power regulation versus general competition law. Criteria for judging competitive versus regulated markets. In F. P. Sioshansi & W. Pfaffenberger (Eds.), *Electricity market reform: An international perspective* (pp. 49–74). Amsterdam: Elsevier.

Knieps, G. (2014). Regulatory reforms of European network industries and the courts. In W. A. Kaal, M. Schmidt, & A. Schwartze (Eds.), *Festschrift zu Ehren von Christian Kirchner: Recht im ökonomischen Kontext* (pp. 917–934). Tübingen: Mohr Siebeck.

Knieps, G., & Zenhäusern, P. (2010). Phasing out sector-specific regulation in European telecommunications. *Journal of Competition Law and Economics, 6*(4), 995–1006.

Knieps, G., & Zenhäusern, P. (2014). Regulatory fallacies in global telecommunications: The case of international mobile roaming. *International Economics and Economic Policy, 11*(1), 63–79.

Köthenbürger, M., Sinn, H.-W., & Whalley, J. (Eds.). (2006). *Privatization experiences in the European Union*. Cambridge, MA: MIT Press.

Newbery, D. M. (2000). *Privatization, restructuring, and regulation of network utilities*. Cambridge, MA: MIT Press.

Niskanen, W. A. (1971). *Bureaucracy and representative government*. Chicago, IL: Aldine.

Owen, B., & Braeutigam, R. (1978). *The regulation game: A strategic use of the administration process*. Cambridge, MA: Ballinger.

Peltzman, S. (1976). Toward a more general theory of regulation. *Journal of Law and Economics 19*, 211–240.

Posner, R. A. (1974). Theories of economic regulation. *Bell Journal of Economics, 5*, 335–358.

Schwarcz, D. (2012). *Preventing capture through consumer empowerment programs: Some evidence from insurance regulation*. Minnesota Legal Studies Research Paper No. 12–06, January 11, Available at SSRN. Retrieved May 20, 2014, from http://papers.ssrn.com/sol3/papers.cfm?abstract_id=1983321.

Spulber, D. F. (1989). *Regulation and markets*. Cambridge, MA: MIT Press.

Spulber, D. F., & Besanko, D. (1992). Delegation, commitment, and the regulatory mandate. *Journal of Law, Economics, and Organization, 8*(1), 126–154.

Stigler, G. J. (1971). The theory of economic regulation. *Bell Journal of Economics, 2*, 3–21.

Stigler, G. J. (1974). Free riders and collective action: An appendix to theories of economic regulation. *Bell Journal of Economics, 5*(2), 359–365.

Varian, H. R. (2010). *Intermediate microeconomics* (8th ed.). New York: Norton.

Vogelsang, I. (1988). Deregulation and privatization in Germany. *Journal of Public Policy, 8*(2), 195–212.

Weingast, B. W., & Moran, M. J. (1983). Bureaucratic discretion or congressional control? Regulatory policymaking by the federal trade commission. *The Journal of Political Economy, 91*(5), 765–800.

Weizsäcker, C. C. von (1982). Staatliche Regulierung – positive und normative Theorie. *Schweizerische Zeitschrift für Volkswirtschaft und Statistik, 3*, 325–343.

Sketch Solutions to the Questions

The objective of the following sketch solutions to the questions is to convey an idea of what is relevant, without providing a step-by-step answer in the sense of a detailed model solution.

Chapter 2

2-1: User Cost of Capital
User cost of capital consists of economic depreciation (depreciation of capital goods) plus the interest (opportunity costs of capital employed). Therefore it should not include any profit components and thus be profit neutral. In accordance with the principle of market reference, when planning decision-relevant costs the developments on the relevant markets should be considered. It is particularly relevant that price and technology developments on the buying and sales markets are taken into account.

2-2: Deprival Value

$$DV(t) = \min\left[RC(t), \max(NRV(t), PV(t))\right]$$

whereby

$RC(t)$ denotes the replacement cost at the beginning of period t,

$NRV(t)$ denotes the net realisable value at the beginning of period t and

$PV(t)$ denotes the net present value at the beginning of period t

The deprival value concept defines the value of a facility by means of the opportunity costs arising when this facility no longer exists. This also takes into account that it can be worthwhile to sell a facility during its economic life or decide not to replace it.

2-3: Economic Depreciation
Closed depreciation schedules assume that at the beginning of a planning period rational expectations regarding the future development of decision-relevant parameters (in particular, prices, technologies) exist. In contrast, open depreciation schedules can take unexpected changes in prices, technologies, etc. into account by means of a sequential adapting of depreciation schedules. In contrast to closed depreciation schedules, open depreciation schedules are unable to guarantee the principle of capital theoretical profit neutrality. However, in network sectors where technological progress and changed demand conditions are particularly relevant, closed depreciation schedules fail to meet the criterion of market reference.

2-4: Path Dependency and Network Evolution
The path dependency of investments describes the fact that a firm will make future investment decisions depending on the entirety of the investments it has already made in the past. One has to differentiate between path dependency on the network infrastructure level and on the network services level. On the network services level, path dependencies are of lesser significance, although timetables for providing network services cannot be changed in the short term either. The real problem is caused by infrastructures, because of the high share of irreversible investments in capital goods involved. Alternative investment strategies (building new networks, expanding existing ones or closing down networks) have to take the path dependency of network infrastructures into account.

2-5: Long-Run Incremental Costs of Novel Network Services
Of central importance is the question whether the network infrastructure used for the provision of higher quality network services has to be exclusively financed by the users of these higher quality network services. The answer to this question depends crucially on the point in time (before or after the decision about a particular network extension) and the expectations on the future development of the network infrastructure. Determining economically well-founded incremental costs is dependent on the entrepreneurial strategy for network evolution in the context of the provision of novel network services. Insofar as the network infrastructure constitutes a multipurpose network which can be utilized for both traditional and novel network services, the costs of the hypothetical novel network infrastructure are not decision-relevant. The situation is different, however, if the variety of network infrastructures is important.

ND
Chapter 3

3-1: Hyper-Congestion

Hyper-congestion does not occur in centrally coordinated network infrastructures, such as railways tracks or airports, but mainly on roads. In contrast to normal congestion, in a hyper-congestion scenario an increase of traffic density does lead to a decrease in traffic flow, because the reduction of speed due to higher traffic density has a greater impact on traffic flow than the increase in traffic flow due to a further increase in the number of vehicles (cf. Fig. 3.2). Because each traffic flow can exist with low as well as with high traffic density, there are cost inefficiencies in case of hyper-congestion. There is a cost correspondence (rather than a cost function), because each possible traffic flow can be achieved by means of two different amounts of costs (with the exception of the capacity limit itself), with the inefficient branch representing the hyper-congestion scenario (cf. Fig. 3.3).

3-2: Braess Paradox

In a scenario with parallel infrastructures of identical quality, there will be a distribution into two equal traffic flows as illustrated in Fig. 3.5. Travelling from a point of origin O to a point of destination D is possible via two parallel paths A and B, as well as one joint path C. On both paths A and B an identical traffic flow occurs spontaneously, so that total travel time on each path is minimized. Incentives for traffic participants to use one path instead of the other do not develop, because after changing from A to B the variable costs of one trip would increase.

The Braess Paradox states that building an intersecting road linking paths A and B (cf. Fig. 3.6) can lead to a situation where the optimal traffic flows described above are no longer incentive compatible, because some individual drivers can work out more time-saving routes, and a slowing down of traffic flows as well as an increase of total travelling time takes place.

If the quality of the route sections is identical, traffic flows will function optimally even after the intersecting road has been built, as there are no incentives for an individual driver to use the intersecting road. The additional time costs of using the intersecting road are not balanced by time saving. Thus the Braess paradox does not occur. The situation is different when, although congestion externalities are identical for the sum of the route sections A and B, the qualities are opposite on the individual sections. By using the intersecting road an individual driver can then choose the two higher quality route sections, resulting in an individual time advantage and a deviation from optimal traffic flows. The Braess paradox occurs.

3-3: Loop Flows

Electric flows disperse through the network in accordance with Kirchhoff's Laws: In every node the sums of incoming and outgoing electricity flows are equal (First Law), and in every closed electric circuit the sum of voltage drops due to the power extracted is equal to the voltage generated by the power fed in (Second Law). Electricity flows on the same path in opposite directions cancel each other out (netting principle). This is best explained by a numerical example (cf. Fig. 3.10). As the path from 1 to 3 via node 2 is twice as long as path (1,3), the flows will split at a ratio of 1 to 2. It is possible to generate 600 MW at entry point 1 and transmit it to exit point 3. Only 400 MW are transmitted via line (1,3); 200 MW are transmitted via line ($z_{1,2} + z_{1,3}$), which is twice as long.

3-4: System Externalities Versus Path-Based Externalities

Electricity transmission networks are fundamentally different from motorways, railway tracks or liquid gas pipelines. In meshed electricity networks it is impossible to transmit electricity on the direct path between an entry point and an exit point. Instead, electricity will choose the path of least resistance through the meshed network. In doing so, at least some amount of electricity will not choose the shortest connection (contract path). The concrete flows on the different lines are not only dependent on the transmission capacities and resistances of the lines, but depend crucially on the feed-in and extraction plans of all entry and exit points. Thus the contract path principle cannot be socially optimal, as pricing according to the linear distance between an entry point and an exit point fails to take the total opportunity costs of network usage into account.

Chapter 4

4-1: Peak Load Pricing

In contrast to the firm peak case, in the shifting peak case, capacity is also fully utilized during the off-peak period. In order to determine the socially optimal total capacity the intersection point of aggregated demand (vertical summation of demand during peak load periods and off-peak periods) with the marginal cost curve is determined. In the shifting peak case, both off-peak demand and peak load demand contribute to the covering of capacity costs (cf. Fig. 4.2 and the resulting optimal prices p_1^o and p_2^o for both periods).

A price differentiation exists, because capacity is fully utilized both during the off-peak period and the peak load period, and different willingness to pay results in different prices for both periods. In contrast, although different prices are charged for both periods in the firm peak case there is no price differentiation. Peak load demand, which is responsible for the marginal costs, pays for them in full.

Sketch Solutions to the Questions

4-2: Two-Part Tariffs
The welfare result of a Pareto improvement through two-part tariffs crucially depends on the option for small customers to stay with the status quo of a linear tariff, if paying a base fee is not worthwhile for them. If this option is not available, there is the risk that small customers will suffer or even leave the market altogether. Thus if the linear price is above marginal costs, the introduction of an optional two-part tariff can mean an improvement for the firm as well as for some consumer groups. At the same time consumers who do not wish to switch to a two-part tariff are not placed in a worse position than before.

4-3: Market Form and Price Differentiation
Price differentiation can exist independent of market form. This is true for both peak load pricing and optional two-part tariffs. The basic principle of peak load pricing in a monopoly is the same as under competition: in the shifting peak case capacity is fully utilized both during the off-peak period and the peak load period. This results in differentiated prices for the peak load versus the off-peak periods. Two-part tariffs leading to Pareto improvements can be found for every linear price, as long as it is not a linear price at marginal costs. Thus the starting point can be either a linear monopoly price or a linear cost-covering price under competition, which, if economies of scale exist, must be higher than marginal costs.

Chapter 5

5-1: Vickrey Auctions
The Vickrey auction is a second-price sealed-bid auction. The highest bidder is awarded the object at a price equal to the second-highest bid. Strategic behaviour is not advantageous, because the bidder runs the risk of either getting the object at a price above his individual willingness to pay, or not getting the object, although his willingness to pay is not exhausted.

The assumption of risk neutrality is irrelevant for the incentive compatibility of the Vickrey auction. In no case will a bidder pay more than his willingness to pay. Strategic bidding below one's individual willingness to pay makes no sense, because it will result in the bidder either not getting the object at all, or getting it at the price of the second highest bid which is not influenced by his bid.

5-2: Auctions and Price Differentiation
Different willingnesses to pay can be exhausted, when more than one unit of an auction object or objects with different product qualities are sold at auction.

5-3: Auctions in Network Industries
As loss-making universal services are not supplied spontaneously under competition, competition for subsidies can be created in the context of auctions. The bidder who offers services of a predefined minimum standard with the lowest subsidy requirements will then win the auction.

5-4: Competitive Invitations to Tender
In London bus transport both individual lines and complete bundles of lines have been auctioned through combinatorial auctions. In doing so it is crucial that the city as commissioning authority for public transport services establishes a binding catalogue of minimal requirements to be provided by the supplier.

Chapter 6

6-1: Network Externalities
Direct network externalities exist if the utility of a network user increases with the number of users active in the network.

Indirect network externalities exist if an increase in the number of consumers of a product increases demand for a complementary product which can then be provided cheaper because of economies of scale in production. For example an increase in the number of consumers of hardware compatible with specific software results in economies of scale in the production of the software.

6-2: Network Externalities and Network Variety
It is necessary that network externalities (network effects) are seen as very important, so that compatibility advantages are considered sufficiently significant for an active participation in the standardization process to be worthwhile. However, the trade-off with network variety can still result in firms building standard coalitions, so that only the technologies of the members of one coalition are compatible.

6-3: Critical Mass
The problem of critical mass states that a minimum number of participants are needed as users of a new network in order to make the network self-sustaining. Government intervention in the form of an active research policy in order to reach critical mass carries the risk of promoting (via subsidies, etc.) the wrong technology. The larger the knowledge problem and the more substantial the network dynamics, the bigger the danger of supporting the wrong technology.

6-4: Committee Solutions
An important function of committees is the search for compromise solutions during the standardisation process. Which solution prevails in a committee is strongly dependent on the membership structure, on the possibility of entry for new members, and on the voting modalities.

Chapter 7

7-1: Internal Subsidisation
Internal subsidisation implies that the incremental cost test is not fulfilled, so that the revenue from at least one service or service bundle does not cover its incremental costs. Thus it holds that:

$$\sum_{i \in S} R_i < \overline{C}(S) \text{ for at least one service bundle}$$

Under the assumption of the the cost covering constraint, the other services must therefore contribute more to the revenue than the costs of their separate production, so that:

$$\sum_{i \in N-S} R_i > C(N-S)$$

This results in incentives for separation and separate production.

7-2: Universal Service Fund
The basic concept of the universal service fund is the creation of a level playing field, both in the profitable and the non-profitable markets. If a sector-specific universal service tax is raised, it is necessary that all providers of profitable services in this sector are included in equal measure in the financing of the fund. The supplier able to offer universal service for the lowest amount of subsidy should be determined in the context of a public tender.

7-3: Case Study Telecommunications
Due to the dynamic developments in the telecommunications sector, politics has considerable room to manoeuvre regarding in particular the scope of universal service. While traditional universal services – for example telephone booths or providing connection to the narrow-band telephone network – become less important, the development of broadband telecommunications networks creates an enormous potential for extending the scope of universal service.

Chapter 8

8-1: Network-Specific Market Power
The network area is either not a natural monopoly, so that several active providers supply the relevant market, or the costs are reversible, that is, not tied to a specific geographic market, so that potential competition can function. Of course, if there is neither a natural monopoly nor irreversible costs, there is no monopolistic bottleneck facility either.

8-2: Potential for Competition on Transport Markets
The precondition for competition on transport markets is that every active and potential supplier is granted equal access conditions to transport infrastructures. Privileged access to scarce infrastructure capacities, for example through excessive prices for competitors, leads to distortions of competition.

8-3: Price Level Regulation of Access Tariffs
On principle, customers should be able to buy at least the same quantity of various services in a given basket of services at the same price level as in the preceding period. *RPI-X* is established as a correction factor, whereby *RPI* constitutes the changes in the consumer price index, and *X* is a percentage to be negotiated between the regulator and the firms, which was in the following interpreted as the percentage of the expected productivity changes in the sector under regulation.

Let $p_{i,t}$ denote the price of the ith product in period t and $q_{i,t-1}$ the quantity of the ith good sold in period t-1, then it follows that the price-cap constraint is:

$$\sum_{i=1}^{n} p_{i,t} \cdot q_{i,t-1} \leq \sum_{i=1}^{n} p_{i,t-1} \cdot q_{i,t-1} \cdot (1 + RPI - X)$$

On the one hand price levels should decrease in favour of consumers, on the other hand recovery of total cost should not be endangered. Price level restrictions should be based on expected productivity changes related to the sector-specific X-factor, and adjusted for inflation. Price structures should not be mandated by regulatory agencies.

Chapter 9

9-1: Normative Versus Positive Theory of Regulation
The normative theory of regulation defines the criteria for determining which network areas should be regulated (regulatory basis) and which instruments should be used to do so (regulatory instruments). The central question is to what extent and how economic sectors should be regulated. In contrast, the

positive theory of regulation studies the emergence, the transformation, and the abolishment, as well as the institutional implementation of sector-specific regulation. Thus the central question is how economic sectors actually are regulated. In examining this question, the influence exercised by individual firms and sectors of the economy, consumer interest, and the bureaucratic self-interest of the regulatory agency must be taken into account. The various interest groups compete with each other in their quest for political influence. The influence exerted by the interest groups is fundamentally dependent on the institutional framework.

The starting point of the normative theory of regulation is the concept of market failure or market imperfections in need of being corrected by regulatory interventions. In contrast, the positive theory of regulation is focussed on the role of interest groups in the regulatory process. Interest groups can exert influence at different stages of the political process, on the legislative procedure as well as on regulatory agencies in their implementation of regulatory laws.

9-2: The Cornerstones of the Regulatory Process
It is necessary to differentiate between legislative institutions, regulatory agencies and the regulated sector. Regulatory legislation typically consists of two components, the legal framework of regulation and the legal regulatory mandate. Within the legal framework of regulation, regulatory agencies possess a certain discretionary freedom of action as regards the implementation of regulation and the choice and application of the regulatory instruments. This leads to the problem of the legal regulatory mandate which restricts the regulatory agency's freedom of action. Depending on the specific legal objectives, appropriate regulatory mandates have to be derived which impose a binding order on the relation between legislator and regulatory agency. Within the framework of the legally specified regulatory mandate, regulation is implemented by the application of the regulatory instruments

9-3: Interest Groups and Regulatory Agencies
The starting point is the question to what extent consumers and producers can assert their respective interests vis-à-vis the regulatory agency. The capture hypothesis assumes that over time regulatory agencies come to be dominated by the regulated industry. The consumer interest groups hypothesis states that regulatory agencies maximise consumer surplus exclusively. Both hypotheses are unsuitable for predicting regulatory agencies' behaviour, because they fail to examine the interaction of the contradictory interests of different interest groups.

9-4: Disaggregated Regulatory Mandate
As regulatory agencies generally cannot be assumed to be maximising social welfare, there is the problem that potential investors expect regulatory opportunism, that is, only successful projects will be compensated. Only by applying

suitable regulatory instruments can over-regulation and double regulation be avoided. The disaggregated regulatory mandate has the objective of reducing the possibilities of opportunistic behaviour by the regulatory agency. It comprises the following basic elements: limiting regulation to areas with network-specific market power, phasing out regulation when network-specific market power no longer exists, and applying incentive regulation in monopolistic bottleneck areas.

Index

A
Access charges, market-compatible, 6
Access holidays, 167
Accounting separation, 152
Airport slots, 83–85
Air traffic control, 9, 118, 139, 141
Air traffic control systems, 118, 140
Air transport, 118, 144
American National Standards Institute (ANSI), 116
Auctions, 7, 87–98
    design, 88, 93, 98
    of frequencies, 97–98
    of third generation mobile communications licenses (UMTS), 98
    types of, 88
Average revenue cap, 151
Awarding rights of way, 87

B
Benchmarking, 153
Braess paradox, 50–52

C
Capacity
    limit, 22, 38–40, 64, 66
    management, 9, 118
    marginal cost function for, 76
    scarcity, 38–40, 83, 84
Capital asset pricing model (CAPM), 20
Capital costs, determination of, 58
Capital theoretical profit neutrality, 12
CAPM. *See* Capital asset pricing model (CAPM)
Capture hypothesis, 161
Chargeable expressways, 59–60

Commissioning of services of general economic interest, 87
Committee solutions, 117
Common costs, 25
    firm-specific, 25–27, 124, 149
    product-group specific, 25–27, 149
Compatibility, 9, 105, 110, 116
Compatibility standards, 9, 101, 106, 110, 117, 118
Compatible network technologies, 105
Competition
    on equal terms rule, 148
    on the infrastructure management level, 139
    between interest groups, 162–165
    on the network infrastructure level, 141–144
    on the network service level, 137
    policy, 7–8
    potential, 94, 135–38, 141, 143, 146
Competitive potentials, disaggregated identification of, 137–144
Competitive strategic behaviour, 8
Complex systems, 2
Computer reservation systems, 8
Concept of the essential facility, 144
Congestion costs, 4, 35–38, 42, 51, 52, 58, 59
Congestion externalities, 7, 35–70
Congestion fees, 35–60, 84
    efficient, 44
    in a monopoly, 54–57
    for quality differentiation, 52
    second-best, 45, 49
    socially optimal, 38, 40–43, 45, 52
    in traffic practice, 57–61
    with variable infrastructure, 42–43
Consumer interest groups hypothesis, 161
Consumption externalities, 101, 102

Converters, 115
Coordination problem, 106, 115–116
Cost-covering constraint, 45–49
Costs
    advantages, 4, 25
    allocation, 22, 26
    of capital, 6, 11–13, 20, 150, 154, 166
        determination of, 11, 21–22
    of efficient service provision, 150
    of electricity generation, 63
    irreversible, 23, 135–137, 143
    long-run avoidable, 22–24
    models, analytical, 29, 149, 150
    strategies, 27–31
Critical mass, 9, 108, 112–116

D

Decision-relevant costs, 6–7, 11, 17, 150
Depreciation
    degressive, 17
    economic, 11, 15–18
    linear, 17
    methods, conventional, 17
    progressive, 17
    schedules
        closed, 18–20
        open, 17–20
Deprival value, 15–18
Deutsches Institut für Normung (DIN), 116
Distribution networks, 1, 4, 154
Dollar votes, 163
Duality principle, 67

E

Economies of scale, 4, 44, 75, 81, 82, 133–138, 146
Economies of scope, 4, 6, 126, 134–38
Efficient Component Pricing Rule (ECPR), 148
Electricity
    networks, 68, 106, 114, 151
    provision of, 5
    sector, 1, 2, 61, 143, 153
Entry point, 4, 61–63, 66, 69
Essential facility doctrine, 144
Euler's theorem, 57
EU Universal Service Directive, 127, 128
Exit point, 4, 61, 69
Expectations, change of, 19
Externalities
    costs, 37, 42, 45, 46, 55, 56, 61, 64
    local, 35
    physical, 35, 64

F

False negative error, 137
False positive error, 137
Financing objectives, 43–49
Fluctuations in capacity utilisation, 13, 22, 40
Frequencies, 87, 97–98, 133
Frequent flyer programmes, 8, 81, 118

G

Gateway technologies, 114
Grandfathering, 93, 95
Graph theory, 3, 49, 50

H

Heavy goods vehicle (HGV), 57–59
Hyper-congestion, 38–40

I

Incentive regulation, 153, 166
Incremental costs, long-run, 22, 23–24, 28–30, 148–150
Infrastructure
    capacities, 3, 61, 137, 138
    management, 3, 9, 118
    policy, 140
    usage fees, 54
Inner city toll, 59
Institutional competition, 88
Interest, 12, 15, 18, 20–21
Interest groups, 157, 161–165
International Standards Organisation (ISO), 116
Internet services, 5
Investment decisions, 28, 40–43, 49, 167
Invitations to tender, 88, 93–97
    of loss-making services, 96
    in public transport, 96–97

J

Joint production, 25

K

Kirchhoff's Laws, 61–63, 65, 66

L

Large technical systems approach, 2
Legal framework of regulation, 158, 166
Legal monopoly, 8

Index

Legal regulatory mandate, 158–160
Legislator, 113, 128, 158–161
Loop flows, 61, 63

M

Marginal costs
  long-run, 11, 22
  short-run, 22
Market power, 8, 121, 133–137, 140, 141, 143, 157, 158, 166
Market power regulation, 7–8, 121, 133–145, 160, 166
Merit order, 65–67
  generalised, 67, 69
Mobile communications networks, 127, 142–144
Mobile telecommunications connections broadband (UMTS), 142
Monopolistic bottleneck, 135–137, 143–47
Multipurpose infrastructures, 28

N

Net present value, 14–17, 29
Net realisable value, 14
Network access technologies, 142
Network development, 28, 109, 150
Network effect, 102–107, 111
Network evolution, 27, 30, 114
Network externalities, 4, 8, 61, 101–106, 109–115, 134, 136
  direct, 101–102
  indirect, 101–102
  physical, 102
  positive, 4
Network industries, 2, 7–9, 29, 93–95, 105, 134
Network infrastructures
  capacities, 3, 7, 21–22, 28, 75, 81, 82, 87, 95
  level of, 3
Network islands, 9, 103, 105, 106, 112–113
Network levels, 3, 21, 95, 106, 134
Network quality, 27
Network services, level of, 3
Network topology, 28, 30
Network usage price, 64
Network variety, 9, 104–105, 114, 116
Newspaper delivery service, 145–147
Non-discriminatory access, 3, 6, 145, 152
Normative theory of regulation, 157

O

Obligation to tender, 87
Off-peak period, 75–78
Open Source Software, 117

Opportunity costs of network utilisation, 4, 7, 61, 64
Optimal price and investment rule, 43, 48
Option value of delayed investments, 29
Overhead costs, 24–26, 148

P

Path dependency, 24, 27–30, 109–110, 116, 150
Peak load period, 75, 76, 78
Peak load pricing, 6, 13, 59, 74, 76–79, 82, 84, 149
  under competition, 76–78
  in a monopoly, 78–79
Phasing out, 123, 142, 147, 152
Pigou–Knight controversy, 52, 54
Pigouvian tax, 52
Positive theory of regulation, 8, 157, 158, 162, 165
Positive theory of the behaviour of regulatory agencies, 158–167
Postal delivery services, 5
Price-cap regulation, 150–154, 160, 166
Price ceilings, 126
Price differentiation, 7, 24, 26, 45, 59, 73–85, 93
  for network services, 81
  for railway tracks, 82–83
  strategies, 7, 73–85
Price level regulation, 94, 150–152
Price structure regulation, 147–150
Public interest theory, 158
Public resources, 3
Public transport services, 5, 96
Public welfare obligations, 122

Q

Quantity discounts, 80

R

Railway infrastructures, 5, 40, 82–83, 143, 144
Railway traffic control, 9, 118, 139–41, 144
Railway traffic control agency, 140
Ramsey prices, 45
Regulation
  demand for, 162
  light-handed, 159
  sector-specific, 8, 136, 147, 157, 160
Regulatory agency, 126, 128, 152, 153, 157–162, 165–167
  freedom of action of, 160–161
  self-commitment capability of, 167
Regulatory approach, disaggregated, 147

Regulatory basis, 157, 159
Regulatory instruments, 8, 147, 157–160
Regulatory mandate, 158–160
  disaggregated, 166–167
Regulatory process, 158, 162
Regulatory rules, 148–149
Regulatory triangle, 159
Rent redistribution, 161–164
Replacement cost, 12, 14–19
Ring networks, 5
Road cost calculation, 58

S
Sector-specific regulation, 133
Separation hypothesis, 150
Services of general economic interest, 122
Slot charges, 84
Specific regulatory intervention, 133
Stand-alone costs, 7, 22, 25, 29–30, 148–149
Standards
  as club goods, 102–104
  as private goods, 102–104
  as public goods, 102–104
Standard-setting, 103, 110, 112–117
Stigler/Peltzman model, 163
Strategies for building a network, 27–29
Subsidisation, internal, 127
Switching costs, 136, 142
Systematic risk (beta factor of the firm), 21
System character of networks, 1–2
System network externalities, 4, 61–69

T
Tariffs, optional two-part, 81
Technical regulatory functions, 9, 117, 133
Technology effect, 102–108
Technology policy, 113–114

Telecommunications, 2, 5, 101, 122, 123, 127–29, 138–144
  networks, 1, 3, 29–31, 141–142
  services, 6, 81, 138, 144
Total costs, 22, 36, 75, 148, 152
Track management, 140
Track price system, 83
Traffic control systems, 140
Traffic density, 38, 39
Traffic flow, 35–43, 49–61
Traffic infrastructure policy, European, 58
Traffic management systems, 139
Traffic services, 137, 140
Transmission networks, 144
Transportation infrastructure, 35–46, 49, 57
Transportation infrastructure capacities, 46, 59

U
UMTS. *See* Mobile telecommunications connections broadband (UMTS)
Universal services, 8, 93, 121–130, 149
  fund, 121, 125–126
    concept of the, 125–126
    expenditures side of the, 126
    financing of, 126
  objectives, 123, 125, 126, 166
  obligations, 121
  programmes, 128
  quality of, 122

W
Water networks, 5
Weighted average cost of capital (WACC), 20

Y
Yield management, 81